MW00949397

Createspace publishing
4900 laCross Road
North Charleston SC, 29406, USA

© 2015 S.Warner, T.Oral, S.Turan

All rights reserved

This book is copyright 2015 with all rights reserved. It is illegal to copy, distribute, or create derivative works from this book in whole or in part or to contribute to the copying, distribution, or creating of derivative works of this book. No part of this book may be stored in a retrieval system or transmitted by any means without the written permission of the author.

Published by Createspace: 03/30/2015

ISBN-13: 978-1511432412
ISBN-10: 1511432411

BOOKS BY DR. STEVE WARNER

28 SAT Math Lessons to Improve Your Score in One Month
    Beginner Course
    Intermediate Course
    Advanced Course

320 SAT Math Problems Arranged by Topic and Difficulty Level
320 SAT Math Subject Test Problems Arranged by Topic and Difficulty Level
    Level 1 Test
    Level 2 Test

SAT Prep Book of Advanced Math Problems
The 32 Most Effective SAT Math Strategies
SAT Prep Official Study Guide Math Companion
SAT Vocabulary Book
320 ACT Math Problems Arranged by Topic and Difficulty Level
320 AP Calculus AB Problems Arranged by Topic and Difficulty Level
320 AP Calculus BC Problems Arranged by Topic and Difficulty Level
555 Math IQ Questions for Middle School Students

BOOKS BY TAYYIP ORAL

555 Math IQ Questions for Middle School Students

555 Math IQ Questions for Elementary School Students
IQ Intelligence Questions for Middle and High School Students
Master's Degree Program Preparation (IQ)

A Text Book for Job Placement Exam in Azerbaijan for Undergraduate and Post Undergraduate Students in Azerbaijan

Algebra (Text Book)

Geometry (Text book)

Geometry Formulas (Text Book)

Algebra Formulas (Text Book)

555 Geometry Problems

for High School Students

135 Questions with Solutions
420 Additional Questions with Answers

Dr. Steve Warner, Tayyip Oral, Serife Turan

© 2015, All Rights Reserved

TABLE OF CONTENTS

Introduction..7

Angles..8

Angles in a Triangle...20

Comparing Sides and Angles in a Triangle.................26

The Pythagorean Theorem and its Converse..............26

Isosceles Right Triangle..30

Perimeter of the Triangle..32

30°, 60°, 90° Triangle...33

Median of a Triangle..34

Angle Bisector of a Triangle...35

Altitude of a Triangle...36

Equilateral Triangle...37

Areas of Triangles..38

Similar Triangles..42

Polygons..47

Regular Polygons...48

Squares..49

Rectangles...53

Rhombus..56

Parallelograms..61

Trapezoids...65

Circles..69

Arc length..70

Areas of Circles and Sectors..76

Rectangular Coordinates...78

Area and Centroid of a Triangle..................................80

Equations of Lines...81

Equations of Circles..88

Rectangular Prisms..91

Cubes...92

Triangular Prisms...93

Pyramids...94

Cylinders...98

Cones 99

Spheres 104

    Test-1 106

    Test-2 109

    Test-3 113

    Test-4 114

    Test-5 115

    Test-6 119

    Test-7 121

    Test-8 122

    Test-9 125

    Test-10 128

    Test-11 130

    Test-12 134

    Test-13 138

    Test-14 141

    Test-15 144

    Test-16 147

    Test-17 148

    Test-18 151

    Test-19 154

    Test-20 158

    Test-21 159

    Test-22 162

    Test-23 166

    Test-24 168

    Test-25 172

    Test-26 176

    Test-27 179

    Test-28 180

Answer Key 181

About the Authors 186

Books by Tayyip Oral 187

Books by Dr. Steve Warner 188

This book was written for high school students, with the goal of developing the basic geometry skills necessary to excel in high school geometry and on standardized tests.

555 Geometry Problems was designed to be used as an additional supplement for a variety of high school geometry courses. The book includes methods, tips, examples with step-by-step solutions, and content tests. The techniques and strategies taught in this book will allow students to solve geometry problems in both conventional and unconventional ways. This will give students several alternative methods when attempting to solve problems. Additionally, the techniques taught here will often allow students to arrive at answers more quickly and to avoid making careless errors.

The practice tests presented in this book are based upon the most recent state level tests and include almost every type of geometry question that one can expect to find on high school level standardized tests. Two assessment tools are provided to measure student progress. The first is 135 example questions with complete explanations, and the second is 28 quizzes to test students more deeply on the content covered. These tools are designed to comfortably build students' test-taking confidence in a natural way.

This book focuses primarily on geometry problems and methods for solving these problems. Definitions and formulas are provided throughout the book. Geometric proofs are not included. We believe that learning geometry will be easier and more fun when this book is used as a resource.

Angles

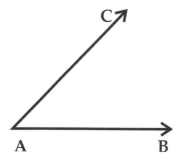

A figure formed by two rays with a common endpoint is called an **angle**.

In the figure above, the angle formed by the rays $\overrightarrow{AB}$ and $\overrightarrow{AC}$ is $\angle BAC$. The point A is called the **vertex** of the angle. The vertex letter is always written between the two others.

The angle above can also be written $\angle CAB$, or it can be abbreviated as $\angle A$ since there is only one angle with vertex A.

Note: $\overrightarrow{AB}$ and $\overrightarrow{AC}$ in the figure above are called **rays** because they have one endpoint. If they had no endpoints they would be called **lines**, and if they had two endpoints they would be called **line segments**.

a)

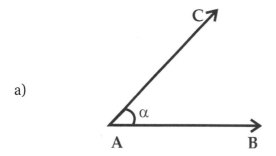

An angle whose measure is between $0°$ and $90°$ is an **acute angle**.

The angle α shown above is acute because $0° < \alpha < 90°$.

Note: We sometimes label an angle with a greek or latin letter such as we did in the figure above with the greek letter α.

b)

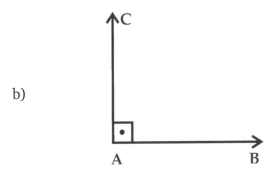

An angle that measures 90° is a **right angle**.

In the figure above we would say that $m\angle A = 90°$ (this is read "the measure of angle A is 90 degrees").

c)

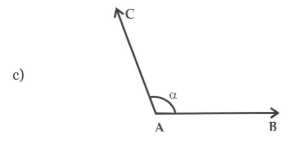

An angle whose measure is between 90° and 180° is an **obtuse angle**.

The angle α shown above is obtuse because $90° < \alpha < 180°$.

d)

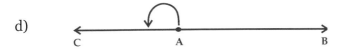

An angle that measures 180° is a **straight angle**.

The angle A shown above is a straight angle because $m\angle A = 180°$.

e)

An angle that measures 360° is a **full angle**.

f)

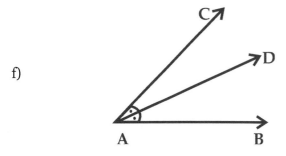

Two angles that have a common vertex and share one common ray, but do not share a common interior, are called **adjacent angles**.

In the figure above, angles $\angle CAD$ and $\angle DAB$ are adjacent angles.

g)

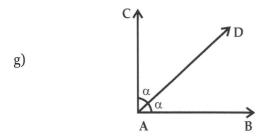

A ray that divides an angle into two angles of equal measure is called an **angle bisector.**

In the figure above ray $\overrightarrow{AD}$ is the angle bisector of $\angle CAB$.

h)

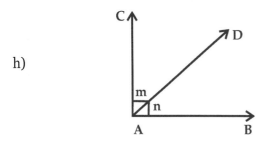

Two angles whose measures sum to 90° are called **complementary angles**.

In the figure above, $\angle CAD$ and $\angle DAB$ are complementary angles because $m\angle CAD + m\angle DAB = 90°$.

Note that in the above figure we could also write $\angle CAD$ as $\angle m$, and we can write $\angle DAB$ as $\angle n$.

i)

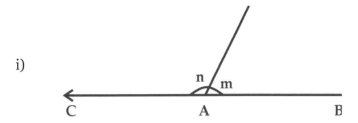

Two angles whose measures sum to 180° are called **supplementary angles**.

In the figure above, $\angle n$ and $\angle m$ are supplementary angles because $m\angle m + m\angle n = 180°$.

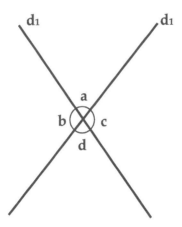

Two non-adjacent angles formed by two intersecting lines are called **vertical angles**.

In the figure above, $\angle a$ and $\angle d$ form a pair of vertical angles, and $\angle b$ and $\angle c$ also form a pair of vertical angles.

Vertical angles are **congruent**, meaning they have the same measure. For example, $m\angle a = m\angle d$. We can also write $\angle a \cong \angle d$. the symbol "$\cong$" can be read "is congruent to."

Note: Sometimes we will just write $a = d$ for short.

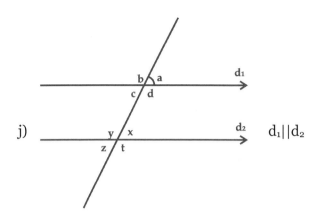

j) $d_1 || d_2$

In the figure above, the two parallel lines d_1 and d_2 are cut by a **transversal** (the unlabeled line).

a and x ⎫
c and z ⎪   **corresponding angles**
b and y ⎬
d and t ⎭

*   Corresponding angles are congruent.

So in the above figure $\angle a \cong \angle x, \angle c \cong \angle z, \angle b \cong \angle y$, and $\angle d \cong \angle t$.

c and x ⎫
d and y ⎬   **alternate interior angles**

*   Alternate interior angles are congruent.

So in the above figure $\angle c \cong \angle x$, and $\angle d \cong \angle y$.

a and z ⎫
b and t ⎬   **alternate exterior angles**

*   Alternate exterior angles are congruent.

So in the above figure $\angle a \cong \angle z$, and $\angle b \cong \angle t$.

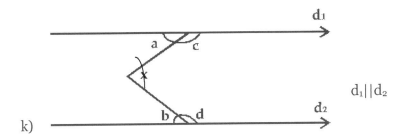

k)

In the figure above, the two parallel lines d_1 and d_2 are cut by two transversals. Note that only part of each of these transversals is drawn.

We have $m\angle x = m\angle a + m\angle b$.

It follows that $m\angle c + m\angle d + m\angle x = 360°$.

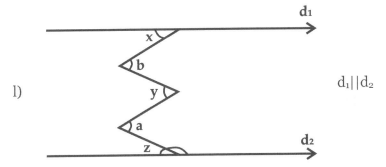

l)

In the figure above, the two parallel lines d_1 and d_2 are cut by four transversals. Note that only part of each of these transversals is drawn.

We have $m\angle x + m\angle y + m\angle z = m\angle a + m\angle b$.

Notational Remark: We may abuse notation in the future by identifying an angle with its measure. For example, we might write the last equation as $x + y + z = a + b$.

In this case we are actually adding the angle measures and not the angles themselves.

Example 1:

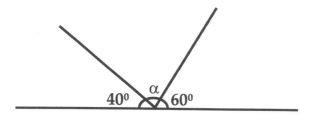

In the figure above, what is the measure of angle α?

Solution:
$$40° + \alpha + 60° = 180°$$
$$\alpha + 100° = 180°$$
$$\alpha = 180° - 100°$$
$$\alpha = \mathbf{80°}$$

Remark: α can be found informally as follows: all three angles shown must have measures that add to 180°. Since 60 and 40 add up to 100, $m\angle\alpha$ must be 180° – 100° = **80°**.

Example 2:

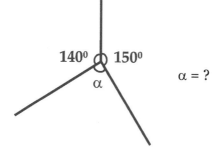

α = ?

Solution:
$$140° + \alpha + 150° = 360°$$
$$290° + \alpha = 360°$$
$$\alpha = 360° - 290°$$
$$\alpha = \mathbf{70°}$$

Remark: α can be found informally as follows: all three angles shown must have measures that add to 360°. Since 140 and 150 add up to 290, $m\angle\alpha$ must be 360° − 290° = **70°**.

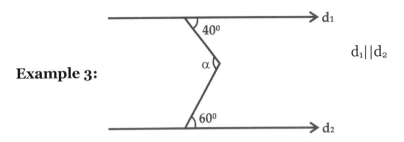

$d_1||d_2$

Example 3:

In the figure above, what is the measure of angle α?

Solution: $\alpha = 40° + 60° = \mathbf{100°}$.

Example 4: The difference between two supplementary angles is 30^0. Find the measure of the smaller angle.

Solution:

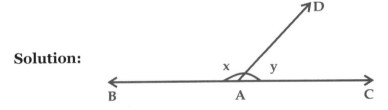

Since x and y are supplementary, $x + y = 180°$. We are also given that $x - y = 30°$. Adding these two equations gives

$$x + y = 180$$
$$\underline{x - y = 30}$$
$$2x = 210$$

So $x = \frac{210}{2} = 105$, and the smaller angle is $y = 180 - 105 = 75°$.

So the measure of the smaller angle is **75°**.

Example 5: Two angles are supplementary and the measure of one is four times the measure of the other. Find the two angle measures.

Solution: The sum of the measures of the supplementary angles is 180^0. If one of the angles is x, the other one is $4x$.

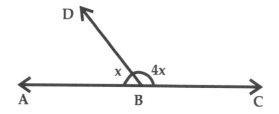

We have $x + 4x = 180$, so that $5x = 180$, and $x = \frac{180}{5} = 36$. It follows that $4x = 4 \cdot 36 = 144$.

So the two angle measures are **36°** and **144°**.

Example 6: Two angles are complementary and the measure of one is twice the measure of the other. Find the two angle measures.

Solution: The sum of the measures of the complementary angles is 90^0. If one of the angles is x, the other one is $2x$.

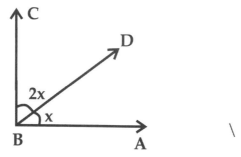

We have $x + 2x = 90$, so that $3x = 90$, and $x = \frac{90}{3} = 30$. It follows that $2x = 2 \cdot 30 = 60$.

So the two angle measures are **30°** and **60°**.

Example 7:

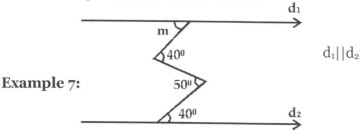

In the figure above, what is the measure of angle m?

Solution: $m + 50 = 40 + 40 = 80$. So $m = 80 - 50 = \mathbf{30°}$.

Example 8:

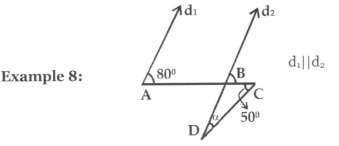

What is the measure of angle α in the figure above?

Solution: $m\angle B = m\angle A = 80°$ because A and B are corresponding angles.

The measure of an exterior angle to a triangle is equal to the sum of the measures of the two opposite interior angles.

It follows that $m\angle B = m\angle C + m\angle D$. So $80 = 50 + \alpha$, and therefore $\alpha = 80 - 50 = \mathbf{30°}$.

Exmaple 9:

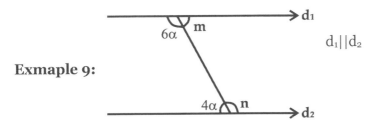

Find α in the figure above.

Solution: Since m and 4α are alternate interior angles , we have $4\alpha + 6\alpha = m + 6\alpha = 180$. So $10\alpha = 180$, and $\alpha = \frac{180}{10} = \mathbf{18°}$.

Example 10: Two angles are supplementary and the measure of one is three times the measure of the other. Find the two angle measures.

Solution: The sum of the measures of the supplementary angles is $180°$. If one of the angles is x, the other one is $3x$.

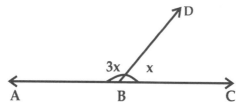

We have $x + 3x = 180$, so that $4x = 180$, and $x = \frac{180}{4} = 45$. It follows that $3x = 3 \cdot 45 = 135$.

So the two angle measures are **45°** and **135°**.

Example 11: The ratio of one angle to another is 5:7. If the two angles are supplementary, find the angles.

Solution: The sum of the measures of the supplementary angles is 180^0. We can represent the two angles by $5x$ and $7x$.

We have $5x + 7x = 180$, so that $12x = 180$, and $x = \frac{180}{12} = 15$. It follows that $5x = 5 \cdot 15 = 75$ and $7x = 7 \cdot 15 = 105$.

So the two angle measures are **75°** and **105°**.

Example 12: Two angles are complementary and the measure of one is three times the measure of the other. Find the two angle measures.

Solution: The sum of the measures of the complementary angles is 90°. If one of the angles is x, the other one is $3x$.

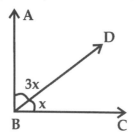

We have $x + 3x = 90$, so that $4x = 90$, and $x = \frac{90}{4} = 22.5$. It follows that $3x = 3 \cdot 22.5 = 67.5$.

So the two angle measures are **22.5°** and **67.5°**.

Example 13: The ratio of one angle to another is 2:3. If the two angles are complementary, find the angles.

Solution: The sum of the measures of the complementary angles is 90°. We can represent the two angles by $2x$ and $3x$.

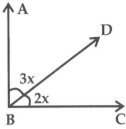

We have $2x + 3x = 90$, so that $5x = 90$, and $x = \frac{90}{5} = 18$. It follows that $2x = 2 \cdot 18 = 36$ and $3x = 3 \cdot 18 = 54$.

So the two angle measures are **36°** and **54°**.

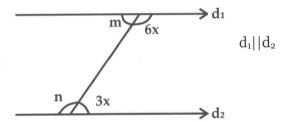

Example 14:

$d_1||d_2$

Find x in the figure above.

Solution: As in example 9, we have $9x = 3x + 6x = 180$ so that $x = \frac{180}{9} = 20°$.

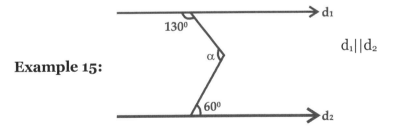

Example 15:

$d_1||d_2$

What is the measure of angle α?

Solution: $\alpha = 60° + 50° = $ **110°**.

Note: $180 - 130 = 50$.

Angles in a Triangle

1)

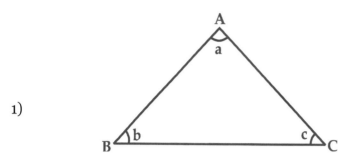

The sum of the measures of the interior angles of a triangle is 180°.
In the figure above, $a + b + c = 180°$.

2)

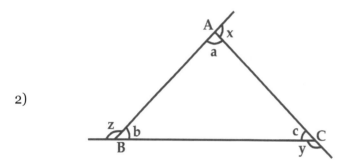

The sum of the measures of the exterior angles of a triangle is 360°.

In the figure above, $x + y + z = 360°$.

3) The measure of an exterior angle to a triangle is equal to the sum of the measures of the two opposite interior angles.

In the figure in 2) we have

$$x = b + c$$

$$y = a + b$$

$$z = a + c$$

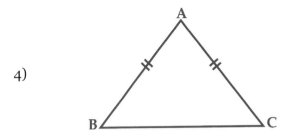

4)

A triangle is **isosceles** if it has two sides of equal length.

In the figure above, the matching tick marks indicate that sides AB and AC have the same length. We write $AB = AC$ in this case.

We can also say that AB and AC are congruent, and write $\overline{AB} \cong \overline{AC}$.

An isosceles triangle also has two angles of equal measure – the angles which are opposite the sides of equal length. So in the figure above we have $m\angle B = m\angle C$, or equivalently $\angle B \cong \angle C$.

In the figure above angles B and C are called **base angles**, and angle A is called the **vertex angle** of the isosceles triangle.

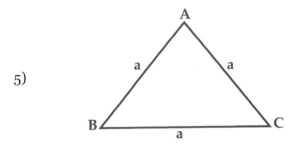

5)

A triangle is **equilateral** if all three sides have equal length.

In the figure above, all three sides have length a, and therefore the triangle is equilateral. We have $AB = BC = AC$.

Equilateral triangles are also **equiangular**. In other words all three angles have equal measure. Since the angles of a triangle have measures that sum to 180°, each angle of an equilateral triangle must measure 60°.

So in the figure above, $m\angle A = m\angle B = m\angle C = 60°$.

6)

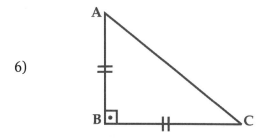

In the figure above we have an **isosceles right triangle**. Note that $AB = BC$ and $m\angle B = 90°$.

Since the angles of a triangle have measures that sum to 180°, and angles A and C are congruent, we have $m\angle A = m\angle C = 45°$.

7)

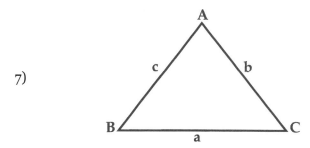

The **perimeter** of a triangle is the sum of the lengths of the three sides of the triangle.

$$P = BC + AC + AB = a + b + c.$$

Example 16: The measure of a base angle of an isosceles triangle is 70°. Find the measure of the vertex angle.

Solution:

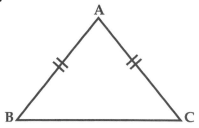

$B + C = 70 + 70 = 140$. Therefore $A = 180 - 140 =$ **40°**.

Notes: (1) Angles B and C are both base angles. They each therefore have measure 70°.

(2) Angle A is the vertex angle.

(3) Since the sum of the measures of the interior angles of a triangle is 180°, we have $A + B + C = 180°$.

Example 17: The interior angles of a triangle are in the ratio 1:3:5. Find the measure of the smallest angle.

Solution: Denote the angles by A, B, and C, and let $A = x$, $B = 3x$, and $C = 5x$. Then we have $x + 3x + 5x = 180$, or equivalently $9x = 180$. So $x = \frac{180}{9} = 20$.

The measure of the smallest angle of the triangle is the measure of angle A which is $x = \mathbf{20°}$.

Example 18: The angle measures of two interior angles of a triangle are 60° and 70°. Find the third angle.

Solution: The third angle measures $180 - 60 - 70 = \mathbf{50°}$.

Example 19: The angle measures of two exterior angles of a triangle are 120° and 140°. Find the third exterior angle.

Solution: The third angle measures $360 - 120 - 140 = \mathbf{100°}$.

Recall: The sum of the measures of the exterior angles of a triangle is 360°.

Example 20:

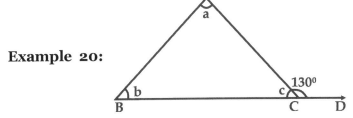

Given that $m\angle A = 50°$ and $m\angle ACD = 130°$, find $m\angle B$.

Solution: $A + B = ACD$. So $50 + B = 130$, and therefore we have $B = 130 - 50 = \mathbf{80°}$.

Recall: The measure of an exterior angle to a triangle is equal to the sum of the measures of the two opposite interior angles.

In problem 20, the exterior angle we are concerned with is $\angle ACD$, and the two opposite interior angles are angles A and B.

Example 21:

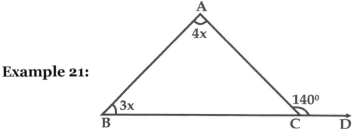

Given that $m\angle A = 4x$, $m\angle B = 3x$, and $m\angle ACD = 140°$, find x.

Solution: $A + B = ACD$. So $4x + 3x = 140$, or equivalently $7x = 140$. So $x = \frac{140}{7} = \mathbf{20°}$.

Note: As in example 20, we used the fact that the measure of an exterior angle to a triangle is equal to the sum of the measures of the two opposite interior angles.

Example 22:

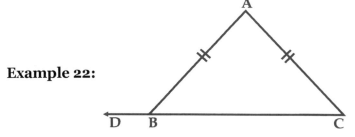

Given that $AB = AC$ and $m\angle ABD = 140°$, find $m\angle A$.

Solution: $ABC = 180 - ABD = 180 - 140 = 40$ because ABD and ABC are supplementary. Since $AB = AC$, $ACB = ABC = 40$, so that $A = 180 - 40 - 40 = \mathbf{100°}$.

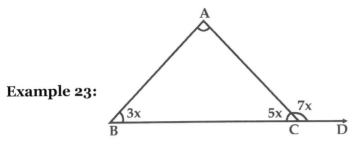

Example 23:

Find $m\angle A$ in the figure above.

Solution: $5x + 7x = 180$, or $12x = 180$, so that $x = \frac{180}{12} = 15$. Therefore $A = 180 - 3x - 5x = 180 - 8x = 180 - 8 \cdot 15 = \mathbf{60°}.$

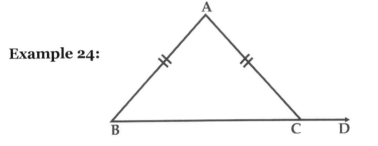

Example 24:

In the isosceles triangle above, exterior angle ACD measures 130°. Find the measure of the vertex angle of the triangle.

Solution: $ACB = 180 - ACD = 180 - 130 = 50$ because ACB and ACD are supplementary. Since $AB = AC$, $ABC = ACB = 50$, so that $A = 180 - 50 - 50 = \mathbf{80°}.$

Comparing Sides and Angles of a Triangle

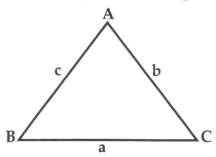

The longest side of a triangle is opposite the angle with largest measure and vice versa. Similarly, the shortest side is opposite the angle of smallest measure.

Symbolically, $a < b < c$ if and only if $m\angle A < m\angle B < m\angle C$.

Triangle Rule: The length of the third side of a triangle is between the difference and sum of the lengths of the other two sides.

For example, if $a < b$ in the figure above, then $b - a < c < b + a$.

The Pythagorean Theorem and its Converse

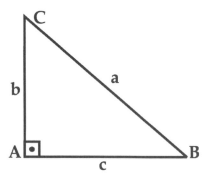

The Pythagorean Theorem says that if a right triangle has a hypotenuse of length a, and legs of lengths b and c, then

$$a^2 = b^2 + c^2.$$

The converse of the Pythagorean Theorem is also true: If a triangle has sides with length a, b, and c satisfying $a^2 = b^2 + c^2$, then the triangle is a right triangle.

More specifically, we have the following.

$a^2 > b^2 + c^2$ if and only if the angle opposite the side of length a is greater than 90 degrees.

$a^2 < b^2 + c^2$ if and only if the angle opposite the side of length a is less than 90 degrees.

Example 25:

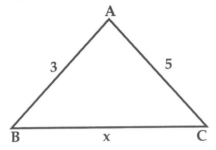

In the figure above, find all possible integer values of x.

Solution: By the triangle rule, $5 - 3 < x < 5 + 3$, or equivalently $2 < x < 8$. The integers between 2 and 8 are **3, 4, 5, 6, 7**.

Example 26:

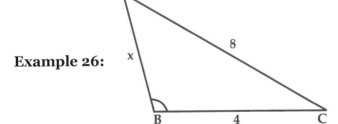

If $\angle B$ is an obtuse angle then find the possible integer values of $x = AB$.

Solution: By the triangle rule, $8 - 4 < x < 8 + 4$, or equivalently $4 < x < 12$. Since $\angle B$ is obtuse, we have $x^2 + 4^2 < 8^2$, or equivalently $x^2 < 64 - 16 = 48$. So $x < \sqrt{48} < 7$. It follows that $4 < x < 7$. The integers between 4 and 7 are **5, 6**.

Example 27:

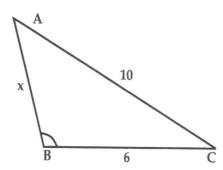

In the figure above $\angle B$ is the largest of the three angles in the triangle. What are all possible integer values for x?

Solution: By the triangle rule, $10 - 6 < x < 10 + 6$, or equivalently $4 < x < 16$. Since $\angle B$ is the largest angle, AC is the longest side. So $x < 10$. It follows that $4 < x < 10$. The integers between 4 and 10 are **5, 6, 7, 8, 9**.

Example 28:

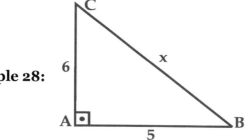

Find x in the figure above.

Solution: By the Pythagorean Theorem,
$$x^2 = 6^2 + 5^2 = 36 + 25 = 61.$$
So $x = \sqrt{61}$.

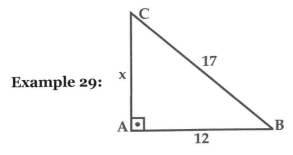

Example 29:

Find x in the figure above.

Solution: By the Pythagorean Theorem, $17^2 = x^2 + 12^2$. So $289 = x^2 + 144$. It follows that $x^2 = 289 - 144 = 145$. Taking the positive square root of each side gives $x = \sqrt{145}$.

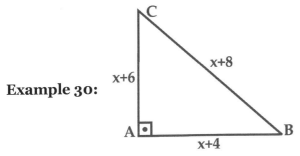

Example 30:

Find the perimeter of the triangle above.

Solution: By the Pythagorean Theorem we have

$$(x + 8)^2 = (x + 4)^2 + (x + 6)^2$$
$$x^2 + 8x + 8x + 64 = x^2 + 4x + 4x + 16 + x^2 + 6x + 6x + 36$$
$$x^2 + 16x + 64 = 2x^2 + 20x + 52$$
$$0 = x^2 + 4x - 12$$
$$0 = (x + 6)(x - 2)$$

So $x = -6$ or $x = 2$

We reject $x = -6$ because this would make AB negative. So $x = 2$. The perimeter of the triangle is

$$(x + 4) + (x + 6) + (x + 8) = 3x + 18 = 3(2) + 18 = 6 + 18 = \mathbf{24}.$$

Isosceles Right Triangle

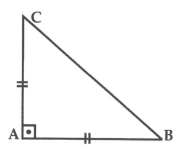

In an isosceles right triangle, both base angles measure 45° and the length of the hypotenuse is the length of either leg times $\sqrt{2}$.

The triangle in the figure above is a right triangle and an isosceles triangle. Therefore $m\angle B = m\angle C = 45°$. If $AB = a$, then it follows that $AC = a$ and $BC = a\sqrt{2}$.

Here are some examples of isosceles right triangles:

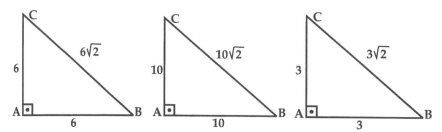

Example 31: The length of the hypotenuse of an isosceles right triangle is $10\sqrt{2}$ cm. Find the perimeter of the triangle.

Solution: Let's draw a picture.

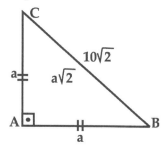

Since the hypotenuse has a length of $10\sqrt{2}$, each leg has length 10. So the perimeter is $10 + 10 + 10\sqrt{2} = \mathbf{20 + 10\sqrt{2}}$ **cm.**

Note: We can also find a by using the Pythagorean Theorem.

$$a^2 + a^2 = \left(10\sqrt{2}\right)^2$$
$$2a^2 = 10^2\left(\sqrt{2}\right)^2$$
$$2a^2 = 100(2)$$
$$a^2 = 100$$
$$a = 10 \text{ cm}$$

Example 32: Find the length of a leg of an isosceles right triangle whose hypotenuse has length 8 cm.

Solution:

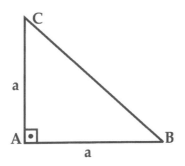

$8 = a\sqrt{2}$ so that $a = \frac{8}{\sqrt{2}} = \frac{8}{\sqrt{2}} \cdot \frac{\sqrt{2}}{\sqrt{2}} = \frac{8\sqrt{2}}{2} = \mathbf{4\sqrt{2}}$ **cm.**

Note: Once again, we can use the Pythagorean Theorem instead.

$$a^2 + a^2 = 8^2$$
$$2a^2 = 64$$
$$a^2 = 32$$
$$a = \sqrt{32} = \sqrt{16 \cdot 2} = \sqrt{16}\sqrt{2} = 4\sqrt{2}.$$

Perimeter of a Triangle

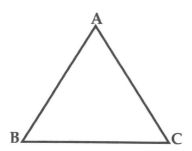

The **perimeter** of a triangle is the sum of the lengths of the three sides of the triangle.

In the figure above, the perimeter of the triangle is $AB + BC + AC$.

Example 33:

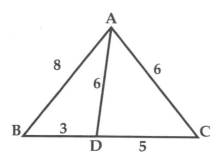

In the figure above, find the ratio of the perimeter of $\triangle ABD$ to the perimeter of $\triangle ACD$.

Solution: $\frac{\text{Perimeter}(ABD)}{\text{Perimeter}(ACD)} = \frac{8+3+6}{6+5+6} = \frac{17}{17} = \mathbf{1}.$

Example 34: The perimeter of a triangle is 60 cm and the sides of the triangle are in the ratio 2:4:6. Find the length of the shortest side of the triangle.

Solution: The sides have length $2x$, $4x$, and $6x$ for some x. Since the perimeter is 60, we have $2x + 4x + 6x = 60$, or equivalently $12x = 60$. The shortest side of the triangle is $2x = \frac{60}{6} = \mathbf{10}$ **cm.**

Note: We could also first solve for x by dividing each side of the equation $12x = 60$ by 12 to get $x = 5$. Just remember that the shortest side is *not* x. It is $2x = 2 \cdot 5 = 10$ cm.

$30°, 60°, 90°$ **Triangle**

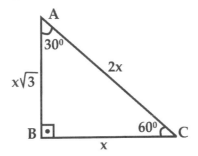

In a triangle with angles $30°$, $60°$, and $90°$, the side opposite the $30°$ angle is half the length of the hypotenuse, and the side opposite the $60°$ angle is $\sqrt{3}$ times half the length of the hypotenuse.

Here are some examples of $30°, 60°, 90°$ triangles:

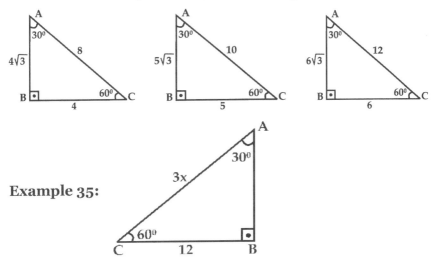

Example 35:

Find x in the figure above.

Solution: Since the triangle is a $30°$, $60°$, $90°$ triangle, we have that $3x = 2 \cdot 12 = 24$. So $x = \frac{24}{3} = $ **8.**

Note: Once again, in a 30°, 60°, 90° triangle the side opposite the 30° angle is half the length of the hypotenuse. Equivalently, the length of the hypotenuse is twice the length of the side opposite the 30° angle. So $3x = 2 \cdot 12$.

Example 36:

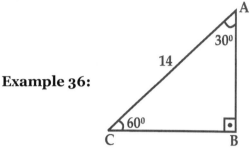

Find the perimeter of the above triangle.

Solution: $BC = \frac{AC}{2} = \frac{14}{2} = 7$ and so $AB = 7\sqrt{3}$. It follows that the perimeter of the triangle is $7 + 7\sqrt{3} + 14 = \mathbf{21 + 7\sqrt{3}}$.

Median of a Triangle

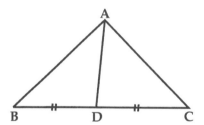

A **median** of a triangle is a line segment from a vertex of the triangle to the opposite side that divides that side into two congruent segments.

In the figure above note that $BD = DC$. It follows that AD is a median of the triangle.

If $\angle A$ is a right angle, then the median AD has the same length as the two congruent segments BD and DC as shown in the picture below.

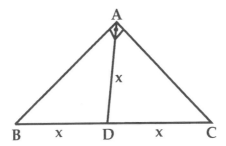

Example 37: Here are some examples. Note that these figures are *not* drawn to scale.

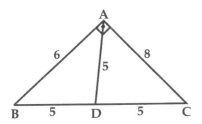 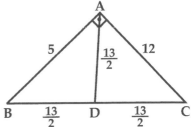

Angle Bisector of a Triangle

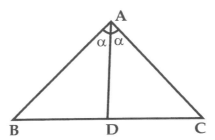

An **angle bisector** of a triangle is a line segment from a vertex of the triangle to the opposite side that divides the angle into two congruent angles.

In the figure above note that $m\angle BAD = m\angle DAC$. It follows that AD is an angle bisector of the triangle.

If AD is an angle bisector of $\triangle ABC$, then $\frac{AB}{BD} = \frac{AC}{DC}$.

Example 38:

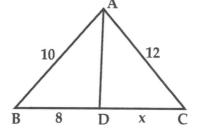

In the figure above, AD is an angle bisector of the triangle. Find x.

Solution: $\frac{AB}{BD} = \frac{AC}{DC}$. So $\frac{10}{8} = \frac{12}{x}$. Therefore $10x = 12 \cdot 8 = 96$. So we have $x = \frac{96}{10} = \mathbf{9.6}$.

Altitude of a Triangle

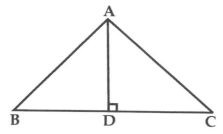

An **altitude** of a triangle is a line segment from a vertex of the triangle to the opposite side that is perpendicular to that side.

In the figure above note that $AD \perp BC$ (read AD is **perpendicular** to BC). It follows that AD is an altitude of the triangle.

Note: In an isosceles triangle, the median, angle bisector and altitude from the vertex angle to the opposite base are all equal.

Equilateral Triangle

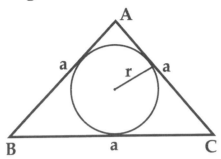

Recall that a triangle is **equilateral** if all three sides have equal length. In the figure above, $AB = BC = AC = a$. Also recall that an equilateral triangle is equiangular so that each angle of the triangle measures 60°.

The **perimeter** of this triangle is $AB + BC + AC = 3a$.

The **area** of this triangle is $A = \frac{a^2\sqrt{3}}{4}$

The circle drawn in the figure above is called the circle **inscribed** inside the triangle. Observe that this circle intersects the triangle in exactly three points.

The length of any altitude of this triangle is $h = 3r$ where r is the radius of the inscribed circle.

The radius r satisfies $r = \frac{a\sqrt{3}}{6}$.

Example 39: The perimeter of an equilateral triangle is 30 cm. Find the area and altitude of the triangle.

Solution: Let a be the length of a side of the triangle. Then we have $3a = 30$, so that $a = \frac{30}{3} = 10$ cm.

The area of the triangle is then $A = \frac{10^2\sqrt{3}}{4} = \mathbf{25\sqrt{3}}$ **cm²**.

The altitude of the triangle is $h = 3r = \frac{3 \cdot 10\sqrt{3}}{6} = \mathbf{5\sqrt{3}}$ **cm.**

Example 40: Find the area of the equilateral triangle with side length 12 cm.

Solution: $Area = \dfrac{12^2\sqrt{3}}{4} = \mathbf{36\sqrt{3}}$ **cm².**

Areas of Triangles

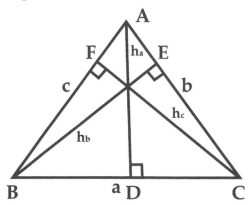

The area of a triangle is equal to half of the product of the length of a side of the triangle and the length of the corresponding altitude.

In the figure above, we have

$$Area = \frac{BC \cdot AD}{2} = \frac{AC \cdot BE}{2} = \frac{AB \cdot CF}{2}$$

or equivalently

$$Area = \frac{a \cdot h_a}{2} = \frac{b \cdot h_b}{2} = \frac{c \cdot h_c}{2}$$

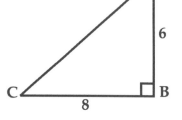

Example 41:

Compute the area of $\triangle ABC$ shown above

Solution: $Area = \frac{1}{2} \cdot 8 \cdot 6 = \textbf{24}.$

Example 42:

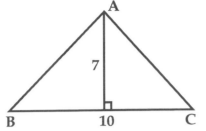

Compute the area of $\triangle ABC$ shown above

Solution: $Area = \frac{1}{2} \cdot 10 \cdot 7 = \textbf{35}.$

Example 43:

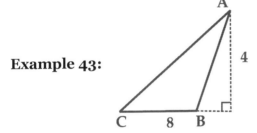

Compute the area of $\triangle ABC$ shown above.

Solution: $Area = \frac{1}{2} \cdot 8 \cdot 4 = \textbf{16}.$

Some Area Rules:

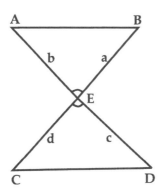

In the figure above, we have

$$\frac{Area(\Delta AEB)}{Area(\Delta CED)} = \frac{a \cdot b}{d \cdot c}$$

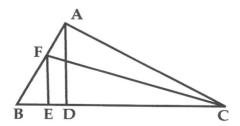

The ratio of the areas of two triangles with equal bases is equal to the ratio of the corresponding altitudes.

In the figure above, we have

$$\frac{Area(\Delta ABC)}{Area(\Delta FBC)} = \frac{AD}{FE}$$

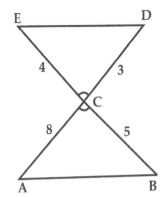

Example 44:

Compute the ratio of the area of ΔECD to the area of ΔACB in the figure above.

Solution: $\dfrac{Area(\Delta ECD)}{Area(\Delta ACB)} = \dfrac{3\cdot4}{8\cdot5} = \dfrac{12}{40} = \dfrac{3}{10}.$

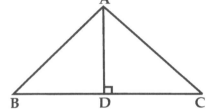

Example 45:

If $BD = 10$, $DC = 12$, and $AD = h$, find $Area(\Delta ABD):Area(\Delta ACD)$.

Solution: $\dfrac{Area(\Delta ABD)}{Area(\Delta ACD)} = \dfrac{BD\cdot AD}{DC\cdot AD} = \dfrac{10h}{12h} = \dfrac{5}{6}.$

Similar Triangles

Two triangles are **similar** if their angles are congruent. Note that similar triangles **do not** have to be the same size.

If triangles ABC and DEF are similar, we write $\triangle ABC \sim \triangle DEF$.

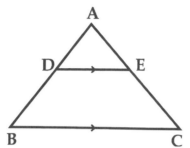

In the figure above, we are given that $DE \parallel BC$. It follows that $\triangle ADE \sim \triangle ABC$.

To see that the two triangles are similar note that (i) they share angle A, (ii) angles ADE and ABC are equal because they are alternate interior angles, and (iii) angles AED and ACB are equal because they are alternate interior angles.

Note that to show that two triangles are similar we need only show that two pairs of angles are congruent. We get the third pair for free because all triangles have 180 degrees.

So in the above figure we only needed to show two of (i), (ii), and (iii) to get similarity.

Note: Corresponding sides of similar triangles are in proportion.

We therefore have $\frac{AD}{AB} = \frac{AE}{AC} = \frac{DE}{BC}$ in the above figure.

In addition we also have $\frac{AD}{BD} = \frac{AE}{EC}$ in the above figure.

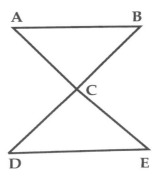

In the figure above, if $AB \parallel DE$, then $\triangle ABC \sim \triangle DEC$.

To see this we can use alternate interior angles as we did for the last figure (also angles ACB and DCE are vertical, thus congruent).

We have $\dfrac{AC}{CE} = \dfrac{BC}{CD} = \dfrac{AB}{DE}$

Example 46:

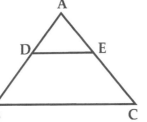

In the figure above, suppose that $DE \parallel BC$, $AD = BD$, $AE = EC$, and $DE = 8$. Find BC.

Solution:

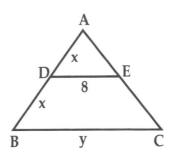

If we let $AD = BD = x$ and $BC = y$, then we have $\dfrac{AD}{AB} = \dfrac{DE}{BC}$, so that $\dfrac{x}{2x} = \dfrac{8}{y}$. Therefore $\dfrac{8}{y} = \dfrac{1}{2}$, and so $BC = y = 8 \cdot 2 = \mathbf{16}$.

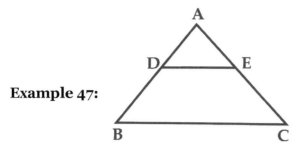

Example 47:

In the figure above, suppose that $DE \parallel BC$, $AD = 6$, $BD = x + 2$, $DE = 4$, and $BC = 2x + 5$. Find x.

Solution: We have $\frac{AD}{AB} = \frac{DE}{BC}$, so that $\frac{6}{6+x+2} = \frac{4}{2x+5}$. Cross multiplication gives $6(2x + 5) = 4(x + 8)$. So $12x + 30 = 4x + 32$ and therefore $8x = 2$. Finally, $x = \frac{2}{8} = \frac{1}{4}$.

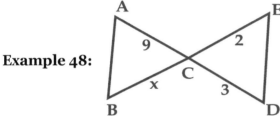

Example 48:

In the figure above, suppose that $AB \parallel DE$, $AC = 9$, $CE = 2$, $BC = x$, and $CD = 3$. Find x.

Solution: $\frac{x}{2} = \frac{9}{3}$, and therefore $3x = 2 \cdot 9 = 18$. So $x = \frac{18}{3} = 6$.

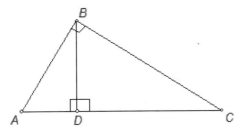

In the figure above we have a right triangle with an altitude drawn from the right angle to the hypotenuse. In this figure triangles BDC, ADB and ABC are similar to each other. Here is what the 3 triangles look like when we draw them separately so that congruent angles match up.

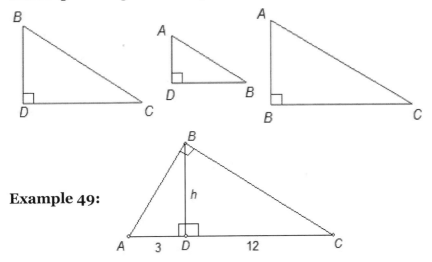

Example 49:

Find h in the figure above.

Solution: We redraw the three triangles next to each other so that congruent angles match up.

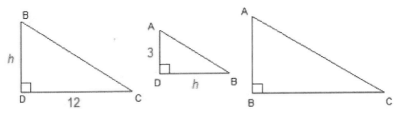

45

We now set up a ratio, cross multiply, and divide: $\frac{h}{12} = \frac{3}{h}$. So $h^2 = 36$, and therefore $h = \mathbf{6}$.

Remark: We didn't really need to redraw the third triangle.

Example 50:

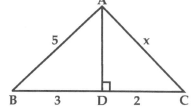

Find x in the figure above.

Solution: We first use the Pythagorean Theorem in $\triangle ABD$ to find AD: $3^2 + AD^2 = 5^2$. So $AD^2 = 25 - 9 = 16$, and therefore $AD = 4$.

We now use the Pythagorean Theorem in $\triangle ACD$ to find x. We have $x^2 = 2^2 + 4^2 = 4 + 16 = 20$. So $x = \sqrt{20} = \sqrt{4 \cdot 5} = \sqrt{4}\sqrt{5} = \mathbf{2\sqrt{5}}$.

Note: For the leftmost triangle, we could have used the Pythagorean triple 3, 4, 5.

In other words, since we know that that the hypotenuse of the triangle has length 5 and one of the legs has length 3, the other leg must have length 4.

In general, Pythagorean triples are sets of three numbers that satisfy the Pythagorean Theorem. The two most common Pythagorean triples are 3,4,5 and 5,12,13.

When using Pythagorean triples remember that the hypotenuse is always the longest side of the triangle.

Example 51:

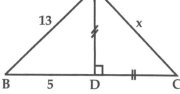

Find x in the figure above.

Solution: $AD^2 + 5^2 = 13^2$. So $AD^2 = 169 - 25 = 144$ and $AD = 12$. So $DC = 12$ also. Finally $x = \mathbf{12\sqrt{2}}$.

Notes: (1) For the leftmost triangle, we could have used the Pythagorean triple 5, 12, 13.

(2) For the rightmost triangle, we can note that $\triangle ADC$ is an isosceles right triangle (45°, 45°, 90° triangle), so that the length of the hypotenuse is $\sqrt{2}$ times the length of a leg of the triangle.

(3) Alternatively, for the rightmost triangle, we can use the Pythagorean Theorem:

$$x^2 = 12^2 + 12^2 = 144 + 144 = 288.$$

So $x = \sqrt{288} = \sqrt{2 \cdot 144} = \sqrt{144}\sqrt{2} = \mathbf{12\sqrt{2}}.$

Polygons

A **polygon** is a two-dimensional geometric figure formed of three or more straight sides.

The sum of the measures of the interior angles of a polygon with n sides is $(n - 2) \cdot 180°$.

The sum of the measures of the exterior angles of a polygon is 360°.

The number of diagonals in an n-sided polygon is $\frac{n(n-3)}{2}$ (a **diagonal** of a polygon is a line segment that connects two vertices of the polygon but does not form an edge of the polygon).

Example 52: Find the sum of the measures of the interior angles of a polygon with five sides.

Solution: $(n - 2) \cdot 180 = (5 - 2) \cdot 180 = 3 \cdot 180 = \mathbf{540°}$.

Remark: A five sided polygon is also called a **pentagon**.

Example 53: Find the number of diagonals in a six sided polygon.

Solution: $\frac{n(n-3)}{2} = \frac{6 \cdot 3}{2} = \mathbf{9}$.

Remark: A six sided polygon is also called a **hexagon**.

Example 54: Find the sum of the measures of the interior angles of a polygon with four times as many diagonals as sides.

Solution: Let n be the number of sides of the polygon. We are given that $\frac{n(n-3)}{2} = 4n$, or equivalently $n(n-3) = 8n$. Dividing each side by n gives $n - 3 = 8$, and so $n = 8 + 3 = 11$.

The sum of the measures of the interior angles of an 11 sided polygon is $(n-2) \cdot 180 = (11-2) \cdot 180 = 9 \cdot 180 = \textbf{1620}°$.

Example 55: The sum of the measures of the interior angles of a polygon is 5 times the sum of the measures of its exterior angles. How many sides does this polygon have?

Solution: $(n-2) \cdot 180 = 5 \cdot 360$, so that $n - 2 = \frac{5 \cdot 360}{180} = 10$. Therefore $n = 10 + 2 = \textbf{12}$.

Regular Polygons

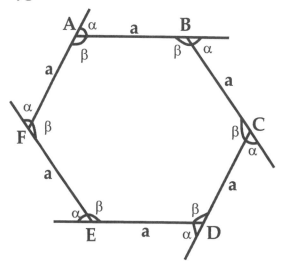

A **regular** polygon is a polygon with all sides equal in length, and all angles equal in measure.

In the figure above, we have a regular hexagon.

$AB = BC = CD = DE = EF = FA = a$, all interior angles have measure β, and all exterior angles have measure α.

The sum of the measures of the interior angles of a hexagon is $(n-2) \cdot 180 = 4 \cdot 180 = 720$. So $\beta = \frac{720}{6} = 120°$ and $\alpha = \frac{360}{6} = 60°$.

In general the measure of an interior angle of a regular polygon with n sides is $\beta = \frac{(n-2)\cdot 180}{n}$, and the measure of an exterior angle of a regular polygon with n sides is $\alpha = \frac{360}{n}$.

Example 56: What is the measure of an interior angle of a regular polygon with an equal number of diagonals and sides.

Solution: We have $\frac{n(n-3)}{2} = n$, so that $n(n-3) = 2n$. So $n - 3 = 2$, and therefore $n = 5$. The sum of the measures of the interior angles is then $(n-2) \cdot 180 = 3 \cdot 180 = 540°$, and so the measure of one interior angle is $\frac{540}{5} = \mathbf{108°}$.

Remark: We can solve the equation $n(n-3) = 2n$ more formally as follows: $n^2 - 3n = 2n$ is equivalent to $n^2 - 5n = 0$. Therefore we have $n(n-5) = 0$, and therefore $n = 0$ or 5.

Squares

A four sided polygon is called a **quadrilateral**.

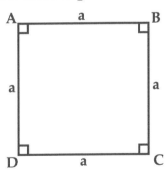

The regular quadrilateral in the figure above is called a **square**.

$m\angle A = m\angle B = m\angle C = m\angle D = 90°$ and $AB = BC = CD = DA = a$.

The perimeter of the square is $4a$ and the area is a^2.

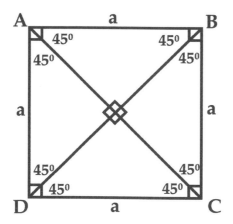

Each diagonal of the square drawn above has length $a\sqrt{2}$. So $AC = BD = a\sqrt{2}$.

Note also that the diagonals are perpendicular and they bisect each of the four interior angles of the square.

So $AC \cong BD$, $AC \perp BD$, AC bisects $\angle A$ and $\angle C$, and BD bisects $\angle B$ and $\angle D$.

Another way to compute the area of a square is Area $= \dfrac{d^2}{2}$ where d is the length of either diagonal of the square.

Also $AC^2 + BD^2 = 4a^2$.

Example 57: Find the area of a square with diagonal of length $8\sqrt{2}$.

Solution: $Area = \dfrac{d^2}{2} = \dfrac{\left(8\sqrt{2}\right)^2}{2} = \dfrac{64 \cdot 2}{2} = \mathbf{64}.$

Example 58: Find the perimeter of a square with a diagonal of length $\sqrt{7}$ cm.

Solution:

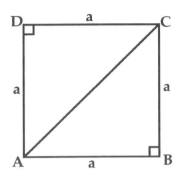

We are given that $AC = \sqrt{7}$. It follows that $a = \frac{\sqrt{7}}{\sqrt{2}} = \frac{\sqrt{7}}{\sqrt{2}} \cdot \frac{\sqrt{2}}{\sqrt{2}} = \frac{\sqrt{14}}{2}$.

So the perimeter of the square is $4a = 4\left(\frac{\sqrt{14}}{2}\right) = \mathbf{2\sqrt{14}}$.

Note: To see that $a = \frac{\sqrt{7}}{\sqrt{2}}$, we can either use the fact that the diagonal of a square has length $\sqrt{2}$ times the length of a side (because $\triangle ABC$ is a $45°, 45°, 90°$ triangle), or we can use the Pythagorean Theorem.

Example 59: Find the area of a square with a perimeter of 3 cm.

Solution: The perimeter of the square is $4a$ where a is the length of a side of the square. So we are given $4a = 3$, so that $a = \frac{3}{4}$. The area of the square is then $a^2 = \left(\frac{3}{4}\right)^2 = \frac{9}{16}$ **cm²**.

Example 60: Find the area and perimeter of a square with a side whose length is 2^x cm.

Solution: Area $= (2^x)^2 = 2^{2x} = (2^2)^x = \mathbf{4^x}$.

Perimeter $= 4(2^x) = 2^2 \cdot 2^x = \mathbf{2^{2+x}}$.

Example 61: Find the perimeter of a square with a diagonal of length $10\sqrt{2}$ cm.

Solution: The length of the diagonal d satisfies $d = a\sqrt{2}$ where a is the length of a side of the square. We are given $d = 10\sqrt{2}$. So $a = 10$, and therefore the perimeter is $4a = 4 \cdot 10 = \textbf{40 cm}$.

Example 62: Find the area of a square with side length $(\sqrt{5 + 2\sqrt{6}})$ cm.

Solution: Area $= \left(\sqrt{5 + 2\sqrt{6}}\right)^2 = \textbf{5} + \textbf{2}\sqrt{\textbf{6}}$ **cm²**.

Example 63:

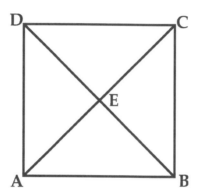

The figure above shows a square with $DE = 6$ cm. Find the area of the square.

Solution: The diagonal DB has length $d = 2DE = 2 \cdot 6 = 12$ cm. So the area of the square is $\dfrac{d^2}{2} = \dfrac{12^2}{2} = \dfrac{144}{2} = \textbf{72 cm²}$.

Rectangles

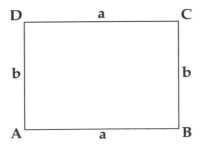

A quadrilateral with four right angles is called a **rectangle**.

$m\angle A = m\angle B = m\angle C = m\angle D = 90°$, $AB = DC = a$, $AD = BC = b$.

The perimeter of the rectangle in the figure above is $2a + 2b = 2(a + b)$ and the area is ab.

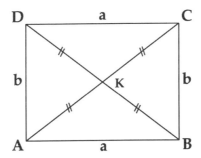

The diagonals of a rectangle are congruent and bisect each other.

So in the figure above we have $AK = KC = BK = KD$.

By the Pythagorean Theorem $AC = BD = \sqrt{a^2 + b^2}$.

Notes: (1) If $a = b$ in the figure above, then the rectangle is a square.

(2) The diagonals of a rectangle are perpendicular if and only if the rectangle is a square.

(3) The diagonals of a rectangle bisect the angles of the rectangle if and only if the rectangle is a square.

Example 64: Find the perimeter, area and length of a diagonal of a rectangle whose sides have lengths 6 cm and 8 cm.

Solution: Perimeter $= 2(a + b) = 2(6 + 8) = 2 \cdot 14 = \textbf{28 cm.}$

Area $= ab = 6 \cdot 8 = \textbf{48 cm}^2.$

Diagonal $= \sqrt{a^2 + b^2} = \sqrt{6^2 + 8^2} = \sqrt{36 + 64} = \sqrt{100} = \textbf{10 cm.}$

Example 65: The area of a rectangle with sides of length x and $5x$ is 80. Find the perimeter of the rectangle.

Solution: We are given $x(5x) = 80$, or equivalently $5x^2 = 80$. So $x^2 = \frac{80}{5} = 16$, and therefore $x = 4$. So the side lengths of the rectangle are 4 and $5 \cdot 4 = 20$, and the perimeter is $2(4 + 20) = \textbf{48}.$

Example 66: Find the perimeter and area of a rectangle with diagonals of length 13 cm and a side of length 5 cm.

Solution:

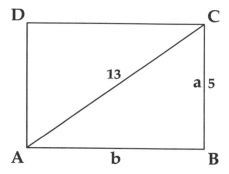

Using the Pythagorean Theorem, or even better, the Pythagorean triple 5, 12, 13, we see that $b = 12$.

Perimeter $= 2(5 + 12) = \textbf{34 cm.}$

Area $= 5 \cdot 12 = \textbf{60 cm}^2.$

Example 67:

The perimeter of a rectangle is 140 cm. The ratio of the side lengths is 3 to 7. Find the area and length of a diagonal of the rectangle.

Solution:

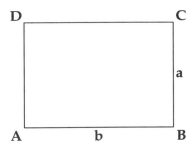

We let $a = 3x$ and $b = 7x$, so that the perimeter of the rectangle is $2(3x + 7x) = 2(a + b) = 140$, or equivalently $20x = 140$. Therefore $x = \frac{140}{20} = 7$. So the side lengths of the rectangle are $a = 3 \cdot 7 = 21$ and $b = 7 \cdot 7 = 49$.

Area $= ab = 21 \cdot 49 = $ **1029 cm²**, and

$$d = \sqrt{21^2 + 49^2} = \sqrt{2842} = 7\sqrt{58} \text{ cm.}$$

Example 68: The perimeter of a rectangle with sides a and b is 48 cm, and $\frac{a}{b} = \frac{1}{3}$. Find the area of the rectangle.

Solution:

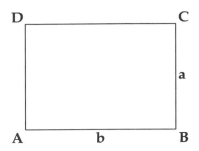

Cross multiplying $\frac{a}{b} = \frac{1}{3}$ gives $b = 3a$. Since the perimeter of the rectangle is 48, we have $48 = 2(a + b) = 2(a + 3a) = 2(4a) = 8a$. So $a = \frac{48}{8} = 6$, and $b = 3a = 3 \cdot 6 = 18$. It follows that the area of the rectangle is $ab = 6 \cdot 18 = $ **108 cm²**.

Example 69: One of the sides of a rectangle with perimeter 50 cm is 5 cm more than the other side. Find the area of the rectangle.

Solution: let a be the length of the shorter side of the rectangle. Then we have $50 = 2(a + a + 5) = 2(2a + 5) = 4a + 10$.

So $4a = 40$, $a = \frac{40}{4} = 10$, and $a + 5 = 10 + 5 = 15$. So the area of the rectangle is $a(a + 5) = 10(15) = \mathbf{150\ cm^2}$.

Rhombuses

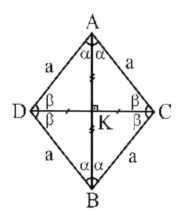

The quadrilateral in the figure above is called a **rhombus**.

All sides of a rhombus have the same length. In the rhombus above, $AC = CB = BD = DA = a$.

Opposite angles of a rhombus are congruent. So $m\angle A = m\angle B$ and $m\angle C = m\angle D$.

The diagonals bisect each other, so that $AK = KB$ and $DK = KC$.

Note also that the diagonals are perpendicular and they bisect each of the four interior angles of the rhombus.

So $AB \perp CD$, AB bisects $\angle A$ and $\angle B$, and CD bisects $\angle C$ and $\angle D$.

The diagonals divide the rhombus into four congruent triangles.

The perimeter of the rhombus is $4a$.

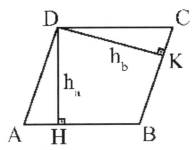

In the figure above, $ABCD$ is a rhombus. DH and DK are altitudes to AB and BC, respectively, and $h_a = h_b$.

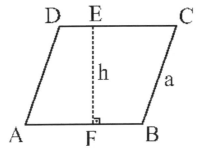

The area of the rhombus in the figure above is ah (base × height).

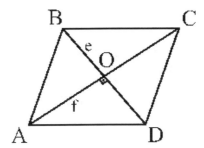

The area of the rhombus in the figure above is $\frac{e \cdot f}{2}$ (the product of the diagonals divided by 2).

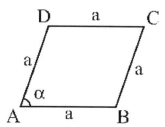

The area of the rhombus in the figure above is $a^2 \sin \alpha$

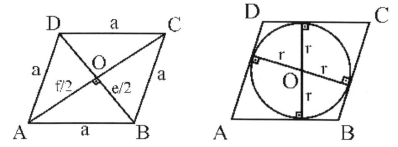

In the leftmost figure above, $e^2 + f^2 = 4a^2$.

In the rightmost figure above, we have inscribed circle O with radius r inside rhombus $ABCD$.

An altitude of the rhombus has length $\frac{e \cdot f}{2a} = 2r$.

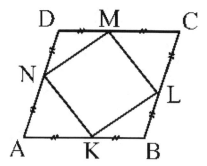

In the figure above, $ABCD$ is a rhombus, and $KLMN$ is the quadrilateral whose vertices are the midpoints of the rhombus. It follows that $KLMN$ is a rectangle.

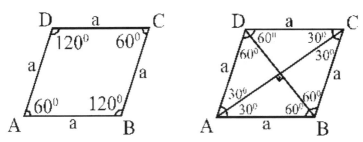

In the leftmost figure above, $ABCD$ is a rhombus, and $m\angle A = 60°$. It follows that $m\angle C = 60°$, and $m\angle B = m\angle D = 120°$.

Additionally, the diagonals form four $30°, 60°, 90°$ triangles, $BD = a$, $AC = a\sqrt{3}$, and the area of the rhombus is $\dfrac{a^2\sqrt{3}}{2}$.

Example 70: Find the perimeter of a rhombus with diagonals of lengths 6 cm and 8 cm.

Solution: $4a^2 = e^2 + f^2 = 6^2 + 8^2 = 100$. So $a^2 = 25$, and therefore $a = 5$. The perimeter of the rhombus is therefore $4a = 4 \cdot 5 = \textbf{20 cm}$.

Example 71: Find the area, perimeter and radius of the inscribed circle of a rhombus with diagonals of lengths 10 cm and 24 cm.

Solution: The area of the rhombus is $\dfrac{e \cdot f}{2} = \dfrac{10 \cdot 24}{2} = \textbf{120 cm}^2$.

Now, $4a^2 = e^2 + f^2 = 10^2 + 24^2 = 676$. So $a^2 = 169$, and $a = 13$. The perimeter of the rhombus is therefore $4a = 4 \cdot 13 = \textbf{52 cm}$.

Finally, $2r = \dfrac{e \cdot f}{2a} = \dfrac{10 \cdot 24}{2 \cdot 13} = \dfrac{120}{13}$. So $r = \dfrac{60}{13}$ **cm.**

Example 72: Find the area, perimeter, altitude and radius of the inscribed circle of a rhombus with diagonals 16 and 30 cm long.

Solution: The area of the rhombus is $\dfrac{e \cdot f}{2} = \dfrac{16 \cdot 30}{2} = \textbf{240 cm}^2$.

Now, $4a^2 = e^2 + f^2 = 16^2 + 30^2 = 1156$. So $a^2 = 289$, and $a = 17$. The perimeter of the rhombus is therefore $4a = 4 \cdot 17 = \textbf{68 cm}$.

$h = \dfrac{e \cdot f}{2a} = \dfrac{16 \cdot 30}{2 \cdot 17} = \dfrac{240}{17}$ **cm.** So $r = \dfrac{h}{2} = \dfrac{120}{17}$ **cm.**

Notes: (1) We can also find a by using the following picture.

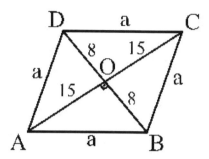

We see that a is the hypotenuse of a right triangle with legs of length 8 and 15. So we have $a^2 = 8^2 + 15^2 = 289$, and therefore $a = \sqrt{289} = 17$.

(2) We can also use the above picture to get the area of the rhombus. Each of the four right triangles has area $\frac{1}{2} \cdot 8 \cdot 15 = 60$. So the area of the rhombus is $4 \cdot 60 = 240$ cm².

Example 73: One of the diagonals of a rhombus is twice the length of the other diagonal. Find the shorter diagonal if the area of rhombus is 100 cm².

Solution: Let x be the length of the shorter diagonal, so that the longer diagonal has length $2x$. Since the area of the rhombus is 100, we have $\frac{x(2x)}{2} = 100$, or equivalently $x^2 = 100$. So the length of the shorter diagonal is $x = \textbf{10 cm}$.

Example 74: One of the diagonals of a rhombus is 3 times longer than the other diagonal. Find the sum of the lengths of the diagonals if the area of the rhombus is 600 cm².

Solution: Let x be the length of the shorter diagonal, so that the longer diagonal has length $3x$. Since the area of the rhombus is 600, we have $\frac{x(3x)}{2} = 600$, or equivalently $x^2 = 400$. So the length of the diagonals are $x = 20$ and $3x = 60$. The sum is then **80 cm.**

Example 75: The side length of a rhombus is $8\sqrt{2}$ cm. Find the area of the rhombus if its smaller angle is 45°.

Solution:

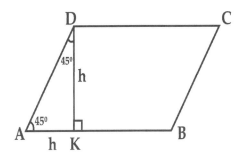

$2h^2 = h^2 + h^2 = a^2 = \left(8\sqrt{2}\right)^2 = 64 \cdot 2 = 128$. So $h^2 = \frac{128}{2} = 64$, and therefore $h = 8$. The area is $ah = \left(8\sqrt{2}\right)(8) = \mathbf{64\sqrt{2}}$ **cm.**

Parallelograms

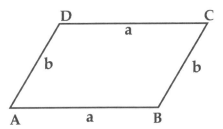

The quadrilateral shown above is a **parallelogram**. The following are equivalent definitions of a parallelogram:

(1) Opposite sides of the quadrilateral are congruent.
(2) Opposite sides of the quadrilateral are parallel.
(3) One pair of opposite sides of the quadrilateral is both congruent and parallel.
(4) Opposite angles of the quadrilateral are congruent.
(5) The diagonals of the quadrilateral bisect each other.

So, in the figure above we have $AB = CD = a$, $AD = BC = b$, $m\angle A = m\angle C$, $m\angle B = m\angle D$, AC and BD (not shown) bisect each other.

We also have $m\angle A + m\angle D = 180°$ and $m\angle B + m\angle C = 180°$ (and therefore also $m\angle A + m\angle B = 180°$ and $m\angle C + m\angle D = 180°$).

The perimeter of the parallelogram is $2a + 2b = 2(a + b)$.

Notes: (1) If $a = b$ in the figure above, then the parallelogram is a rhombus.

(2) If $m\angle A = m\angle D$, then the parallelogram is a rectangle.

(3) If $a = b$ and $m\angle A = m\angle D$, then the parallelogram is a square.

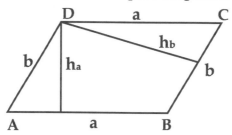

The area of the parallelogram in the figure above is $ah_a = bh_b$. In other words, $Area = base \times height$

The area is also equal to $ab \sin A$.

Note: $\sin A = \sin B = \sin C = \sin D$ so any of the four angles can be used in the last area formula.

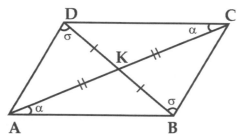

In the parallelogram above, $AK = KC$ and $BK = KD$.

We also have $m\angle CAB = m\angle ACD = \alpha$ and $m\angle ADB = m\angle DBC = \sigma$.

Letting $e = AC$ and $f = BD$, we have $e^2 + f^2 = 2(a^2 + b^2)$.

Note that four congruent triangles are formed by the diagonals, so that all of these triangles have equal area.

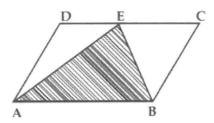

In the figure above, the area of $\triangle AEB$ is half the area of parallelogram $ABCD$.

Example 76: The base of a parallelogram is 12 cm and the corresponding height is 6 cm. Find the area.

Solution: Area = $12 \cdot 6 = $ **72 cm.**

Example 77:

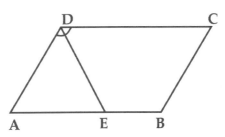

In the figure above, $ABCD$ is a parallelogram, DE bisects $\angle ADC$, $EB = 4$ cm, and $BC = 12$ cm. Find the perimeter of $ABCD$.

Solution:

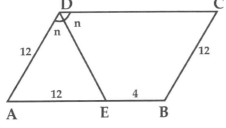

$AD = BC = 12$. Letting $m\angle D = 2n$, we have $m\angle AED = m\angle EDC = n$. So $\triangle ADE$ is isosceles with $AE = AD = 12$. It follows that $AB = AE + EB = 12 + 4 = 16$. So the perimeter of the parallelogram is $2(16 + 12) = $ **56 cm.**

Example 78: In parallelogram $ABCD$, $AD = 12$, $AB = 18$, $m\angle A = 30°$, and $\sin 30° = \frac{1}{2}$. Compute the area of $ABCD$.

Solution:

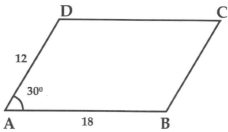

The area of the parallelogram is

$$ab \sin A = 12 \cdot 18 \sin 30° = 12 \cdot 18 \cdot \frac{1}{2} = \textbf{108}.$$

Example 79:

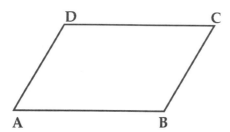

In the figure above, $ABCD$ is a parallelogram, $m\angle A = 60°$, and $m\angle D = 4x$. Find x.

Solution: $4x = m\angle D = 180 - m\angle A = 180 - 60 = 120$.

So $x = \frac{120}{4} = \textbf{30°}$.

Example 80:

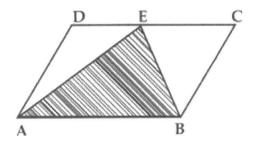

In the figure above, $ABCD$ is a parallelogram with area 40 cm². Find the area of $\triangle AEB$.

Solution: The area of the triangle is half the area of the parallelogram. So the area of $\triangle AEB$ is $\frac{40}{2}$ = **20 cm²**.

Trapezoids

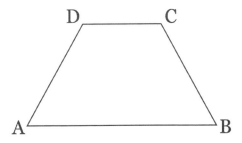

A **trapezoid** is a quadrilateral with one pair of parallel sides. In the figure above, $AB \parallel DC$.

Note that it is not necessary for $AD = BC$ in order for the quadrilateral to be a trapezoid, but if $AD = BC$, then we have an **isosceles trapezoid**. In this case we also have $m\angle A = m\angle B$.

In the figure above, AB and DC are the **bases** of the trapezoid.

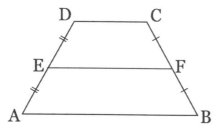

In the trapezoid above, $AE = ED$ and $BF = FC$. In this case, EF is called the **median** (or **midline**) of the trapezoid, and $EF = \frac{AB+DC}{2}$.

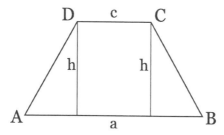

The figure above shows a trapezoid with bases a and c and height h. The area of this trapezoid is

$$Area = (\frac{a+c}{2}) \cdot h$$

In other words, we take the average of the bases and multiply by the height.

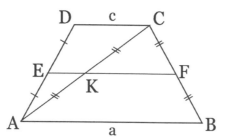

In the trapezoid shown above, the midline EF bisects the diagonal AC. That is $AK = KC$.

We also have $EK = \frac{DC}{2} = \frac{c}{2}$ and $KF = \frac{AB}{2} = \frac{a}{2}$.

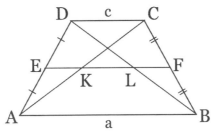

In the trapezoid shown above, $AK = KC$, $BL = LD$, and $KL = \frac{a-c}{2}$.

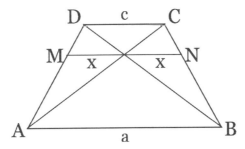

In the trapezoid shown above, $AB \parallel MN \parallel DC$. In this case we have

$$\frac{1}{x} = \frac{1}{a} + \frac{1}{c}$$

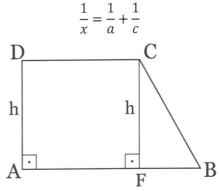

Trapezoid $ABCD$ shown above is a **right trapezoid** because it has two right angles (angles A and D).

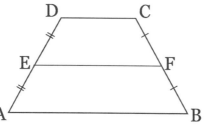

Example 81:

In the trapezoid shown above, EF is the median, $AB = 12$, and $DC = 8$. Find EF.

Solution: $EF = \frac{AB+DC}{2} = \frac{12+8}{2} = \frac{20}{2} = \mathbf{10}$.

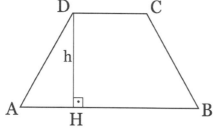

Example 82:

In trapezoid $ABCD$ shown above, $AB = 20$ cm, $DC = 12$ cm, and $DH = 6$ cm. Find the area of $ABCD$.

Solution: Area $= \frac{AB+DC}{2} \cdot DH = \frac{20+12}{2} \cdot 6 = \frac{32}{2} \cdot 6 = 16 \cdot 6 = \mathbf{96\ cm^2}$

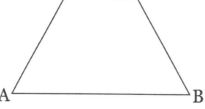

Example 83:

In trapezoid $ABCD$ shown above, $AB = 40$ cm, $DC = 20$ cm, and $AD = BC = 26$ cm. Find the area of $ABCD$.

Solution: Let's add some information to the picture.

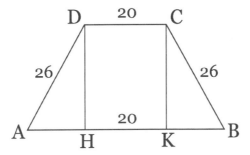

We have $AH = KB = \frac{40-20}{2} = 10$. So $DH = CK = \sqrt{26^2 - 10^2} = 24$.

It follows that the area of $ABCD$ is $\frac{40+20}{2} \cdot 24 = \textbf{720 cm}^{\textbf{2}}$.

Circles

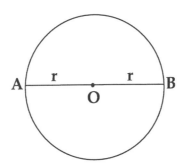

A **circle** is a two-dimensional geometric figure formed of a curved line surrounding a center point, every point of the line being an equal distance from the center point. This distance is called the **radius** of the circle. The **diameter** of a circle is the distance between any two points on the circle that pass through the center of the circle.

In the circle shown above, OA and OB are both radii, and AB is a diameter. Note that $OA = OB = r$ and $AB = 2r$.

The **circumference** of the circle is $C = 2\pi r$, or equivalently $C = \pi d$

π is an irrational number which can be defined as the ratio of the circumference of a circle to its diameter. 3.14 and $\frac{22}{7}$ are close approximations of π.

Arc Length

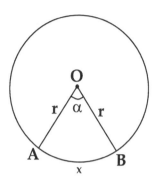

The length of arc AB shown in the figure above is $x = 2\pi r \cdot \frac{\alpha}{360}$, where α is $m\angle AOB$ in degrees.

We will always take the word "arc" to mean the minor arc. This is the smaller of the two arcs that connect A to B.

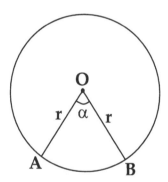

The figure above shows a **central angle**.

The measure of a central angle is equal to the measure of its intercepted arc. So $m\angle\alpha = m\widehat{AB}$.

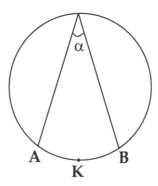

The figure above shows an **inscribed angle**.

The measure of an inscribed angle is equal to half the measure of its intercepted arc. So $m\angle\alpha = \frac{1}{2} \cdot m\overset{\frown}{AKB}$.

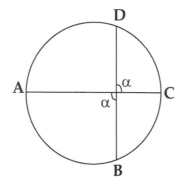

In the figure above, $m\angle\alpha = \frac{1}{2}(m\overset{\frown}{AB} + m\overset{\frown}{CD})$.

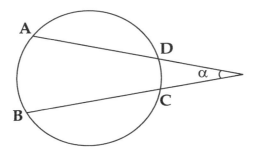

In the figure above, $m\angle\alpha = \frac{1}{2}(m\overset{\frown}{AB} - m\overset{\frown}{CD})$.

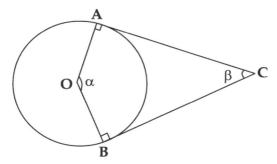

In the figure above, $m\angle\alpha + m\angle\beta = 180°$.

Example 84: Find the circumference of a circle with a diameter of 4 cm.

Solution: $C = \pi d = \pi(4) = \mathbf{4\pi}$ **cm**.

Example 85: Find the radius of a circle with circumference 22π cm.

Solution: $C = 2\pi r$, so that $2\pi r = 22\pi$. So $r = \frac{22\pi}{2\pi} = \mathbf{11}$ **cm**.

Example 86: Find the angle measure of an arc of a circle whose central angle measures 60°.

Solution:

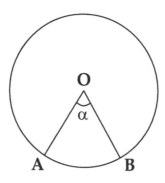

$m\widehat{AB} = m\angle\alpha = \mathbf{60°}$.

Example 87:

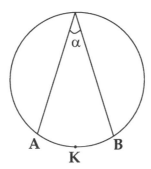

In the figure above, $m\widehat{AKB} = 80°$. Find $m\angle\alpha$.

Solution: $m\angle\alpha = \frac{1}{2}m\widehat{AKB} = \frac{1}{2} \cdot 80 = \mathbf{40°}$.

Example 88:

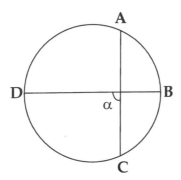

In the figure above, $m\widehat{AB} = 80°$, and $m\widehat{DC} = 100°$. Find $m\angle\alpha$.

Solution: $m\angle\alpha = \frac{1}{2}\left(m\widehat{AB} + m\widehat{CD}\right) = \frac{1}{2}(80 + 100) = \frac{1}{2} \cdot 180 = \mathbf{90°}$.

Example 89:

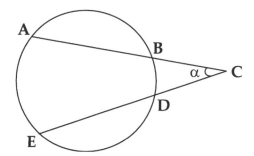

In the figure above, $m\widehat{AE} = 120°$, and $m\widehat{BD} = 40°$. Find $m\angle\alpha$.

Solution: $m\angle\alpha = \frac{1}{2}\left(m\widehat{AE} - m\widehat{BD}\right) = \frac{1}{2}(120 - 40) = \frac{1}{2} \cdot 80 = \mathbf{40°}$.

Example 90:

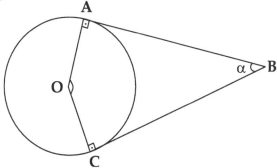

In the figure above, $OA \perp AB$, $OC \perp BC$, and $m\angle AOC = 130°$. Find $m\angle\alpha$.

Solution: $m\angle AOC + m\angle ABC = 180°$. So $130° + m\angle ABC = 180°$, and therefore $m\angle\alpha = m\angle ABC = 180 - 130 = \mathbf{50°}$.

Example 91:

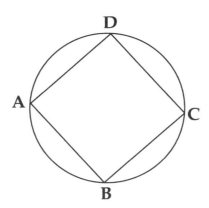

In the figure above, $m\angle A = 4x$ and $m\angle C = 5x$. Find x.

Solution: $m\angle A + m\angle C = \frac{1}{2}BCD + \frac{1}{2}DAB = \frac{1}{2} \cdot 360 = 180°$. So we have $4x + 5x = 180$, so that $9x = 180$, and therefore $x = \frac{180}{9} = \mathbf{20°.}$

Example 92: The radius of a circle is 8. Find the length of an arc of the circle intercepted by a central angle measuring 120°.

Solution:

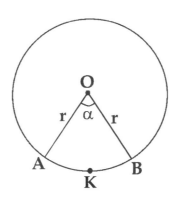

$$2\pi r \cdot \frac{\alpha}{360} = 2\pi \cdot 8 \cdot \frac{120}{360} = \mathbf{\frac{16\pi}{3}.}$$

Areas of Circles and Sectors

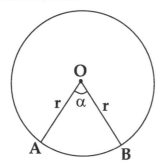

The area of a circle with radius r is $A = \pi r^2$.

The area of sector AOB shown in the figure above is $x = \pi r^2 \cdot \dfrac{\alpha}{360}$, where α is $m\angle AOB$ in degrees.

Example 93: Find the area of a circle with diameter 6 cm.

Solution: $r = \dfrac{d}{2} = \dfrac{6}{2} = 3$. So $A = \pi r^2 = \pi(3^2) = \textbf{9}\boldsymbol{\pi}$ **cm²**.

Example 94: Find the area of a circle with radius $2\sqrt{3}$ cm.

Solution: $A = \pi r^2 = \pi(2\sqrt{3})^2 = \pi(4 \cdot 3) = \textbf{12}\boldsymbol{\pi}$ **cm²**.

Example 95: Find the diameter of a circle with area 144π cm².

Solution: We are given $A = 144\pi$, so that $\pi r^2 = 144\pi$. Therefore $r^2 = 144$, and so $r = 12$. We then have $d = 2r = 2 \cdot 12 = \textbf{24 cm}$.

Example 96: Find the radius of a circle with area π cm².

Solution: $\pi r^2 = \pi$. So $r^2 = 1$, and therefore $r = \textbf{1 cm}$.

Example 97:

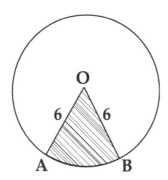

In the figure above $r = 6$ cm, and $m\angle AOB = 60°$. Find the area of the shaded region.

Solution: $\pi r^2 \cdot \frac{\alpha}{360} = \pi(6^2) \cdot \frac{60}{360} = \mathbf{6\pi}$ **cm²**.

Example 98:

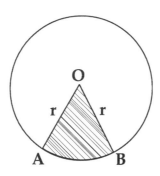

In the figure above $r = 4$ cm, and $m\angle AOB = 90°$. Find the area of the shaded region.

Solution: $\pi r^2 \cdot \frac{\alpha}{360} = \pi(4^2) \cdot \frac{90}{360} = \mathbf{4\pi}$ **cm²**.

Rectangular Coordinates

We form the **rectangular** (or **Cartesian**) **coordinate system** by intersecting a vertical line with a horizontal line. The horizontal line is called the **x-axis**, and the vertical line is called the **y-axis**. The point of intersection of these two lines is called the **origin**, and it has coordinates (0,0).

A point has the form (x, y). The real number x is called the **x-coordinate** of the point and the real number y is called the **y-coordinate** of the point. To plot the point (x, y) we move right or left $|x|$ units from the origin and then up or down $|y|$ units from the origin, depending upon if x and y are positive or negative.

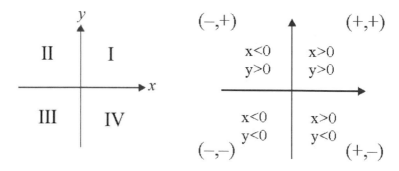

The x- and y-axes split the **xy-plane** into four quadrants. These quadrants are labelled in the figure above on the left. The figure on the right shows which coordinates are positive and negative in each quadrant. For example, in quadrant I, both the x- and y-coordinates of each point are positive, whereas in quadrant II, the x-coordinate of each point is negative, and the y-coordinate of each point is positive.

The **distance** between the points $A(x_1, y_1)$ and $B(x_2, y_2)$ in the plane is

$$d(A, B) = \sqrt{(x_2 - x_1)^2 + (y_2 - y_1)^2}$$

The following figure shows a geometric justification for this formula.

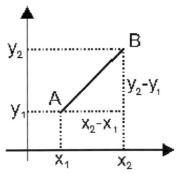

Note that the distance between A and B is simply the length of the line segment AB. By the Pythagorean Theorem, we see that

$$AB^2 = (x_2 - x_1)^2 + (y_2 - y_1)^2.$$

The distance formula follows from this observation.

The distance between the points $A(x_1, y_1, z_1)$ and $B(x_2, y_2, z_2)$ in space is

$$d(A, B) = \sqrt{(x_2 - x_1)^2 + (y_2 - y_1)^2 + (z_2 - z_1)^2}$$

Example 99: Find the distance between the points $A(-1, 7)$ and $B(3, 4)$.

Solution: $d(A, B) = \sqrt{\left(3 - (-1)\right)^2 + (4 - 7)^2} = \sqrt{4^2 + (-3)^2}$

$$= \sqrt{16 + 9} = \sqrt{25} = 5.$$

The **midpoint** of the line segment with endpoints $A(x_1, y_1)$ and $B(x_2, y_2)$ is

$$M = (\frac{x_1 + x_2}{2}, \frac{y_1 + y_2}{2})$$

Note that to get the x-coordinate of the midpoint we are simply taking the arithmetic mean of the x-coordinates of the two points. Similarly, to get the y-coordinate of the midpoint we are taking the arithmetic mean of the y-coordinates of the two points.

Area and Centroid of a Triangle

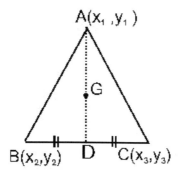

In the figure above, $A(x_1, y_1)$, $B(x_2, y_2)$, and $C(x_3, y_3)$ are vertices of $\triangle ABC$, G is the **centroid** (or **center of gravity**) of the triangle, and D is the midpoint of side BC. We have

$$D = \left(\frac{x_2 + x_3}{2}, \frac{y_2 + y_3}{2}\right) \qquad G = \left(\frac{x_1 + x_2 + x_3}{3}, \frac{y_1 + y_2 + y_3}{3}\right)$$

$$\text{Area of } \triangle ABC = \frac{1}{2}[(x_1 y_2 + x_2 y_3 + x_3 y_1) - (x_1 y_3 + x_3 y_2 + x_2 y_1)]$$

The following figure may help to visualize this area formula:

$$\begin{vmatrix} x_1 & y_1 \\ x_2 & y_2 \\ x_3 & y_3 \\ x_1 & y_1 \end{vmatrix}$$

Example 100: Find the centroid and area of the triangle with vertices $A(7,2)$, $B(1,3)$, and $C(4,1)$.

Solution: The centroid is $\left(\frac{7+1+4}{3}, \frac{2+3+1}{3}\right) = \left(\frac{12}{3}, \frac{6}{3}\right) = (\mathbf{4, 2})$.

The area is $\frac{1}{2}(21 + 1 + 8 - 7 - 12 - 2) = \frac{9}{2}$.

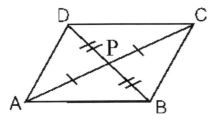

In parallelogram $ABCD$ above, $P(x_0, y_0)$ is the intersection of diagonals AC and BD where A, B, C, D have coordinates $A(x_1, y_1)$, $B(x_2, y_2)$, $C(x_3, y_3)$, and $D(x_4, y_4)$.

$$x_0 = \frac{x_1 + x_3}{2} = \frac{x_2 + x_4}{2} \qquad y_0 = \frac{y_1 + y_3}{2} = \frac{y_2 + y_4}{2}$$

Equations of lines

The **slope** of the line passing through the points $A(x_1, y_1)$ and $B(x_2, y_2)$ is $m = \frac{y_2 - y_1}{x_2 - x_1}$.

Example 101: Find the slope of the line passing through the points $A(2,4)$ and $B(-4,8)$.

Solution: $m = \frac{y_2 - y_1}{x_2 - x_1} = \frac{8-4}{-4-2} = \frac{4}{-6} = -\frac{2}{3}$.

The **point-slope** form of the equation of a line is

$$y - y_1 = m(x - x_1)$$

Where m is the slope of the line and (x_1, y_1) is any point on the line.

If the line passes through the origin $(0,0)$, then the point-slope form gives $y = mx$.

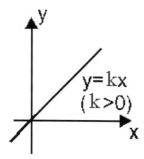

In the figure above, we have a line passing through the origin with slope $m = k > 0$. If $k < 0$ the line would be moving downward from left to right.

Example 102: Sketch the graph of the line with equation $y = 2x$.

Solution:

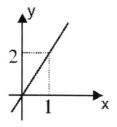

The line passes through $(0,0)$ and the slope of the line is $m = 2 = \frac{2}{1}$. So we start at $(0,0)$ and then move up 2 and right 1 to plot a second point. Once we have two points we can draw the line.

Example 103: Sketch the graph of the line with equation $y = x$.

Solution: $m = 1 = \frac{1}{1}$ (up 1, right 1).

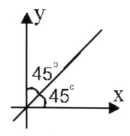

Example 104: Sketch the graph of the line with equation $y = -x$.

Solution: $m = -1 = \frac{-1}{1}$ (down 1, right 1).

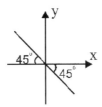

Example 105: Write an equation of the line with a slope of 2 that passes through the point $(2, -3)$.

Solution: An equation in point-slope form is $y - (-3) = 2(x - 2)$, or equivalently $\boldsymbol{y + 3 = 2(x - 2)}$.

Note: If we solve the equation for y we get the **slope-intercept** form for the equation of a line: $y = mx + b$.

Here m is the slope, and the point $(0, b)$ is the y-intercept of the line.

In the last example we can distribute the 2 on the right to get $y + 3 = 2x - 4$. Adding 3 to each side of the equation we get $y = 2x - 7$.

Example 106: Write an equation of the line passing through the points $(2,3)$ and $(-2,1)$.

Solution: $m = \frac{1-3}{-2-2} = \frac{-2}{-4} = \frac{1}{2}$. So, using the point $(2,3)$, an equation of the line is $\boldsymbol{y - 3 = \frac{1}{2}(x - 2)}$.

Notes: (1) If we use the point $(-2,1)$ instead, we get the equation $\boldsymbol{y - 1 = \frac{1}{2}(x + 2)}$.

(2) Solving either of the equations in point-slope form for y gives the same slope-intercept form: $\boldsymbol{y = \frac{1}{2}x + 2}$.

(3) As another alternative, we have $\frac{y-3}{1-3} = \frac{x-2}{-2-2}$, or equivalently $\frac{y-3}{-2} = \frac{x-2}{-4}$. Cross multiplying gives $-2x + 4 = -4y + 12$, or equivalently $\boldsymbol{-2x + 4y = 8}$.

(4) The equation of a line in **general form** is $ax + by = c$ where a, b, and c are real numbers with $a \neq 0$ or $b \neq 0$.

The equation in note (3) is written in general form.

The **two-intercept** form of the equation of a line is

$$\frac{x}{a} + \frac{y}{b} = 1$$

where $(a, 0)$ is the x-intercept and $(0, b)$ is the y-intercept.

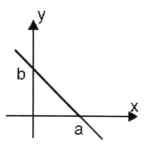

Example 107: Write an equation of the line passing through the points (0,4) and (3,0).

Solution:

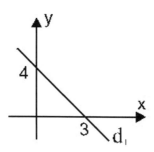

Using the two-intercept form of the equation of the line we get $\frac{x}{3} + \frac{y}{4} = 1$, or equivalently $4x + 3y = 12$.

Example 108: Write an equation of the line passing through the points (0,5) and (−4,0).

Solution:

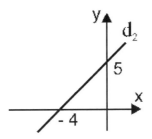

Using the two-intercept form of the equation of the line we get $\frac{x}{-4} + \frac{y}{5} = 1$, or equivalently $-5x + 4y = 20$.

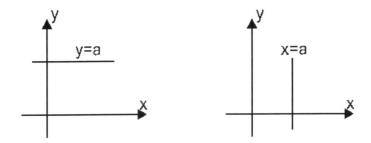

A **horizontal line** has the equation $y = a$ for some real number a, and a **vertical line** has the equation $x = a$ for some real a.

Example 109: Write an equation of the line passing through the points (3,2) and (3,5).

Solution: Both points have 3 as their x-coordinate. The line is therefore a vertical line with equation $x = 3$.

Note: If we were to try to compute the slope of the line we would get $m = \frac{5-2}{3-3} = \frac{3}{0}$ which is undefined. Vertical lines have an **undefined slope**.

Parallel lines have the same slope and perpendicular lines have slopes that are negative reciprocals of each other.

Example 110: Write an equation of the line passing through the point (1,5) and parallel to the line with equation $2y + 3x = 1$

Solution: We solve the given equation for y. We first subtract $3x$ from each side of the equation to get $2y = -3x + 1$. We then divide by 2 to get $y = -\frac{3}{2}x + \frac{1}{2}$.

It is now easy to see that the slope of the given line is $m = -\frac{3}{2}$. Since parallel lines have the same slope, we can now write an equation in point-slope form: $y - 5 = -\frac{3}{2}(x - 1)$.

Example 111: Write an equation of the line passing through the point $(0,3)$ and perpendicular to the line with equation $4y - 5x = 8$

Solution: We solve the given equation for y. We first add $5x$ to each side of the equation to get $4y = 5x + 8$. We then divide by 4 to get $y = \frac{5}{4}x + 2$.

It is now easy to see that the slope of the given line is $m = \frac{5}{4}$. Since perpendicular lines have negative reciprocal slopes, we can now write an equation in slope-intercept form: $y = -\frac{4}{5}x + 3$.

The **general form of an equation of a line** is $ax + by = c$ where a, b and c are real numbers. If $b \neq 0$, then the slope of this line is $m = -\frac{a}{b}$. If $b = 0$, then the line is vertical and has no slope. Let us consider 2 such equations.

$$ax + by = c$$
$$dx + ey = f$$

(1) If there is a number r such that $ra = d$, $rb = e$, and $rc = f$, then the two equations represent the **same line**. Equivalently, the two equations represent the same line if $\frac{a}{d} = \frac{b}{e} = \frac{c}{f}$. In this case the system of equations has **infinitely many solutions**.

(2) If there is a number r such that $ra = d$, $rb = e$, but $rc \neq f$, then the two equations represent **parallel** but distinct lines. Equivalently, the two equations represent parallel but distinct lines if $\frac{a}{d} = \frac{b}{e} \neq \frac{c}{f}$. In this case the system of equations has **no solution**.

(3) Otherwise the two lines intersect in a single point. In this case $\frac{a}{d} \neq \frac{b}{e}$, and the system of equations has a **unique solution**. These three cases are illustrated in the figure below.

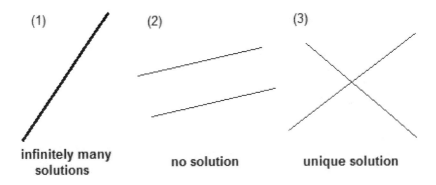

(1) infinitely many solutions

(2) no solution

(3) unique solution

Example 112: The following two equations represent the same line.

$$2x + 8y = 6$$
$$3x + 12y = 9$$

To see this note that $\frac{2}{3} = \frac{8}{12} = \frac{6}{9}$ (or equivalently, let $r = \frac{3}{2}$ and note that $\left(\frac{3}{2}\right)(2) = 3$, $\left(\frac{3}{2}\right)(8) = 12$, and $\left(\frac{3}{2}\right)(6) = 9$).

The following two equations represent parallel but distinct lines.

$$2x + 8y = 6$$
$$3x + 12y = 10$$

This time $\frac{2}{3} = \frac{8}{12} \neq \frac{6}{10}$.

The following two equations represent a pair of intersecting lines.

$$2x + 8y = 6$$
$$3x + 10y = 9$$

This time $\frac{2}{3} \neq \frac{8}{10}$.

Equations of Circles

The **standard form** for the equation of a circle is

$$(x - a)^2 + (y - b)^2 = r^2$$

Where (a, b) is the center, and r is the radius of the circle.

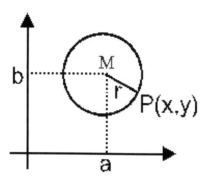

The figure above shows a circle with center $M(a, b)$ and radius r. Note how every point $P(x, y)$ on the circle is at a distance of r from the center.

Example 113: Write an equation of the circle with center $(4,3)$ and radius 5 cm.

Solution: $(x - 4)^2 + (y - 3)^2 = 5^2$, or $(x - 4)^2 + (y - 3)^2 = 25$.

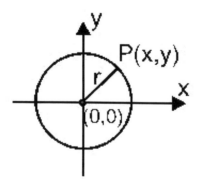

The figure above shows a circle centered at the origin. $(0,0)$. In this case its equation simplifies to $x^2 + y^2 = r^2$.

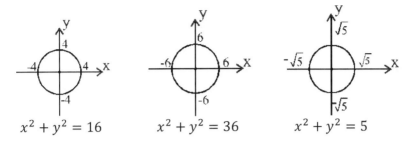

$$x^2 + y^2 = 16 \qquad x^2 + y^2 = 36 \qquad x^2 + y^2 = 5$$

Below are circles with centers on the y-axis and x-axis.

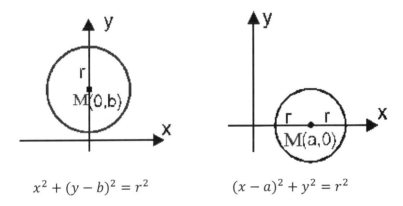

$$x^2 + (y - b)^2 = r^2 \qquad\qquad (x - a)^2 + y^2 = r^2$$

Let's put the equation of a circle into another form:

$$(x - a)^2 + (y - b)^2 = r^2$$
$$x^2 - 2ax + a^2 + y^2 - 2by + b^2 = r^2$$
$$x^2 + y^2 - 2ax - 2by + a^2 + b^2 - r^2 = 0$$

Letting $A = -2a$, $B = -2b$, and $C = a^2 + b^2 - r^2$ gives the **general form** for the equation of a circle.

$$x^2 + y^2 + Ax + By + C = 0$$

Example 114: Find the center and radius of the circle with equation $x^2 + y^2 - 2x + 4y - 4 = 0$.

Solution: We use a procedure called **completing the square**.

First we regroup as follows: $x^2 - 2x + y^2 + 4y = 4$.

Second we half and square the coefficients of x and y: $\left(-\frac{2}{2}\right)^2 = 1$ and $\left(\frac{4}{2}\right)^2 = 4$.

Third we add these numbers to each side of the equation:

$$(x^2 - 2x + 1) + (y^2 + 4y + 4) = 4 + 1 + 4$$

Finally, we factor each expression in parentheses, and add the numbers on the right hand side: $(x - 1)^2 + (y + 2)^2 = 9$.

We now have an equation in standard form, and we see that the center of the circle is $(1, -2)$ and the radius is **3**.

Example 115: Find the radius of the circle with equation $x^2 + y^2 - 4y - 12 = 0$.

Solution: Completing the square gives $x^2 + y^2 - 4y + 4 = 12 + 4$, or equivalently $x^2 + (y - 2)^2 = 16$. So the radius of the circle is **4**.

It should be noted that the equation

$$(x - a)^2 + (y - b)^2 = k$$

may or may not have a graph which is a circle.

If $k > 0$, then the graph is a circle with center (a, b) and radius $\sqrt{k}$.

If $k = 0$, then the graph is the single point (a, b).

If $k < 0$, then the equation has no graph.

Rectangular Prisms

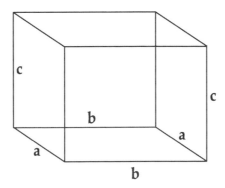

In the figure above we have a **rectangular prism**. The **base** of the prism has sides of length a and b, and the **height** of the prism is c.

The **surface area** of the prism is $A = 2(ab + ac + bc)$, and the **volume** of the prism is $V = abc$.

Example 116: Find the surface area and volume of a rectangular prism with sides of length 6, 7, and 4 cm.

Solution: $A = 2(6 \cdot 7 + 6 \cdot 4 + 7 \cdot 4) = 2 \cdot 94 = \textbf{188 cm}^2$.

$V = 6 \cdot 7 \cdot 4 = \textbf{168 cm}^3$.

Let's look at a rectangular prism with a square base.

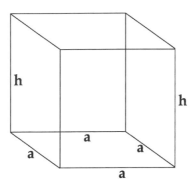

For this special case, we see that the surface area is

$$A = 2(a^2 + ah + ah) = 2a^2 + 4ah$$

and the volume is $V = a^2h$.

Example 117: Find the surface area and volume of a rectangular prism if the base is square with edge length 6 cm, and the height is 4 cm.

Solution: $A = 2 \cdot 6^2 + 4 \cdot 6 \cdot 4 = 72 + 96 = \textbf{168 cm}^\textbf{2}$.

$V = 6^2 \cdot 4 = \textbf{144 cm}^\textbf{3}$.

Cubes

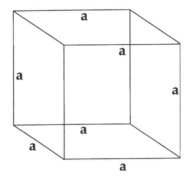

A **cube** is a rectangular prism with all sides equal in length

For this special case, we see that the surface area is

$$A = 2(a^2 + a^2 + a^2) = 6a^2$$

and the volume is $V = a^3$.

Example 118: Find the surface area and volume of a cube with edge length 10 cm.

Solution: $A = 6 \cdot 10^2 = \textbf{600 cm}^\textbf{2}$ and $V = 10^3 = \textbf{1000 cm}^\textbf{3}$.

Triangular Prisms

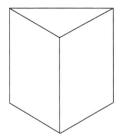

In the figure above we have a **triangular prism**. The two **base faces** of the prism are triangles and the three **lateral faces** are rectangles.

The **surface area** of the prism is $A = 2B + R_1 + R_2 + R_3$, where B is the area of a base triangle and R_1, R_2, and R_3 are the areas of the lateral rectangles.

The **volume** of the prism is $V = Bh$ where B is the area of a base triangle and h is the height of any lateral rectangle.

Example: 119:

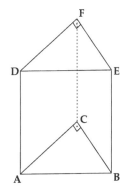

The figure above is a triangular prism with $DF = 12$, $FE = 16$, and $AD = 18$ cm. Find the surface area and volume of the prism.

Solution: The area of the triangle is $B = \frac{1}{2} \cdot 12 \cdot 16 = 96$. By the Pythagorean Theorem, we have that $DE = \sqrt{12^2 + 16^2} = 20$.

Now, we also have $R_1 = 12 \cdot 18 = 216$, $R_2 = 16 \cdot 18 = 288$, and $R_3 = 20 \cdot 18 = 360$.

The surface area of the prism is then

$A = 2B + R_1 + R_2 + R_3 = 2 \cdot 96 + 216 + 288 + 360 = \mathbf{1056 \ cm^2}$.

The volume is $V = Bh = 96 \cdot 18 = \mathbf{1728 \ cm^3}$.

Remarks: (1) We could find DE faster by noticing that $12 = 3 \cdot 4$ and $16 = 4 \cdot 4$, so that $DE = 5 \cdot 4 = 20$. In other words 12, 16, 20 is a multiple of the Pythagorean triple 3, 4, 5.

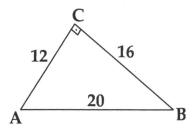

(2) $R_1 + R_2 + R_3 = Ph$ where P is the perimeter of the triangle and h is the height of any lateral rectangle. In this case, we have

$$Ph = (12 + 16 + 20) \cdot 18 = 48 \cdot 18 = 864.$$

So the surface area is $2B + Ph = 2 \cdot 96 + 864 = 1056$ cm².

Pyramids

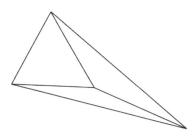

A **pyramid** consists of a base which is a polygon, and triangular sides which meet at a common **vertex**.

In the figure above we have a **triangular pyramid** (because the base is a triangle).

The **surface area** of a pyramid is $A = B + L$, where B is the area of the base, and L is the sum of the areas of the lateral faces.

For example, the surface area of the triangular pyramid shown above is the sum of the base area and the areas of the three lateral faces.

Example 120: Find the surface area of a triangular pyramid if the area of its base is 14 cm² and each of its lateral faces has area 8 cm².

Solution: $A = B + L = 14 + 3 \cdot 8 = 14 + 24 = $ **38 cm².**

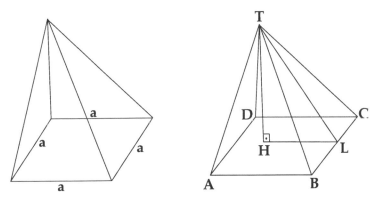

In the figures above we have a **square pyramid** (because the base is a square). In this case $B = a^2$ and there are 4 lateral faces.

TL in the rightmost figure above is called the **slant height** of the pyramid

If, in the rightmost figure, we let $h = TH$, and $AB = a$, then $HL = \frac{a}{2}$. By the Pythagorean Theorem, the slant height is

$$TL = \sqrt{\left(\frac{a}{2}\right)^2 + h^2} = \sqrt{\frac{1}{4}(a^2 + 4h^2)} = \frac{1}{2}\sqrt{a^2 + 4h^2}$$

It follows that the lateral surface area is

$$L = 4 \cdot \frac{1}{2}a \cdot \frac{1}{2}\sqrt{a^2 + 4h^2} = a\sqrt{a^2 + 4h^2}$$

* **Note:** We will only be looking at **right pyramids**. In a right pyramid the vertex is over the midpoint of the base. A pyramid which is not right is called an **oblique pyramid**. So technically our definition of a square pyramid is really that of a right square pyramid.

Example 121: Find the base area of a square pyramid whose base has a side length of 8 cm.

Solution: $B = a^2 = 8^2 = 64$ **cm².**

Example: 122:

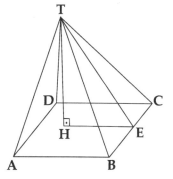

In the figure above, $AB = 10$ cm and $TH = 12$ cm. Find TE.

Solution: First note that $HE = \frac{1}{2}AB = \frac{1}{2} \cdot 10 = 5$. Using the Pythagorean triple 5, 12, 13, we see that $TE = 13$ cm.

Notes: (1) If you don't remember the Pythagorean triple 5, 12, 13, you can also use the Pythagorean Theorem.

(2) TH is the height and TE is the slant height of the pyramid.

Example 123: Find the surface area of a square pyramid with height 2 cm whose base has a side length of 4 cm.

Solution: $B = a^2 = 4^2 = 16$ cm², and

$$L = a\sqrt{a^2 + 4h^2} = 4\sqrt{4^2 + 4 \cdot 2^2} = 4\sqrt{32} = 4 \cdot 4\sqrt{2} = 16\sqrt{2} \text{ cm}^2.$$

So $A = B + L = 16 + 16\sqrt{2} = \mathbf{16(1 + \sqrt{2})}$ **cm².**

Note: Instead of memorizing the formula for L, we can first find $TE = \sqrt{2^2 + 2^2} = \sqrt{8} = \sqrt{4}\sqrt{2} = 2\sqrt{2}$, so that the area of one lateral face is $\frac{1}{2} \cdot 4 \cdot 2\sqrt{2} = 4\sqrt{2}$. Therefore $L = 4 \cdot 4\sqrt{2} = 16\sqrt{2}$ cm².

Example 124: Find the length of a lateral edge of a square pyramid with height 8 cm whose base has a side length of 12 cm.

Solution:

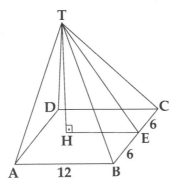

Since $HE = \frac{1}{2} \cdot 12 = 6 = 2 \cdot 3$ and $TH = h = 8 = 2 \cdot 4$, it follows that the slant height is $TE = 2 \cdot 5 = 10$. Using the Pythagorean Theorem, $TB^2 = BE^2 + TE^2 = 6^2 + 10^2 = 36 + 100 = 136$. So a lateral edge has length $TB = \sqrt{136} = \mathbf{2\sqrt{34}}$ **cm**.

The volume of a pyramid is $V = \frac{1}{3}Bh$.

Example 125: Find the volume of a pyramid with an equilateral triangular base of side length 5 cm, and a height of 8cm.

Solution:

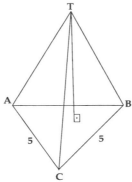

Recall that the area of an equilateral triangle with side length a is $A = \frac{\sqrt{3}}{4}a^2$. So $B = \frac{\sqrt{3}}{4} \cdot 5^2 = \frac{25\sqrt{3}}{4}$. So the volume is

$$V = \frac{1}{3}Bh = \frac{1}{3} \cdot \frac{25\sqrt{3}}{4} \cdot 8 = \frac{50\sqrt{3}}{3} \text{ cm}^3.$$

Cylinders

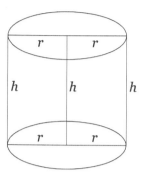

In the figure above we have a **right circular cylinder**. h is the **altitude** (or **height**) of the cylinder and r is the **radius** of the base.

The **surface area** of the cylinder is $A = 2B + L = 2\pi r^2 + 2\pi rh$, where $B = \pi r^2$ is the area of the base and $L = 2\pi rh$ is the **lateral area**.

The **volume** of the cylinder is $V = \pi r^2 h$.

* **Note:** Whenever we mention a cylinder in this book we will always mean a right circular cylinder.

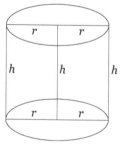

Example 126:

The figure drawn above is a cylinder with base $r = 3$ cm and $h = 8$ cm. Find the surface area and volume of the cylinder.

Solution: $A = 2\pi r^2 + 2\pi rh = 18\pi + 48\pi = \mathbf{66\pi}$ **cm².**

$V = \pi r^2 h = \pi \cdot 9 \cdot 8 = \mathbf{72\pi}$ **cm³.**

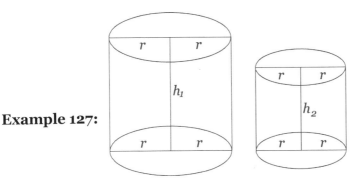

Example 127:

The figures drawn above are cylinders with heights $h_1 = 12$ cm and $h_2 = 8$ cm. If both cylinders have the same base radius, compute the ratio of the volume of the larger cylinder to the volume of the smaller cylinder.

Solution: $\dfrac{V_1}{V_2} = \dfrac{\pi r^2 h_1}{\pi r^2 h_2} = \dfrac{12}{8} = \dfrac{3}{2}$.

Cones

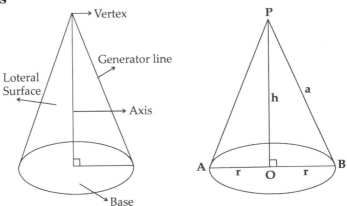

A **right circular cone** is formed when we rotate a right triangle $360°$ about one of its legs.

In the figure above, h is the **altitude** (or **height**) of the cone, a is the **slant height**, and r is the **radius** of the base.

By the Pythagorean theorem we have $a^2 = h^2 + r^2$.

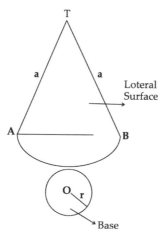

The **surface area** of the cone is $A = B + L = \pi r^2 + \pi r a$, where $B = \pi r^2$ is the area of the base and $L = \pi r a$ is the **lateral area**.

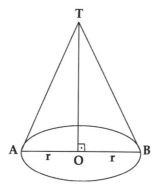

The **volume** of the cone is $= \frac{1}{3}\pi r^2 h$.

* **Note:** Whenever we mention a cone in this book we will always mean a right circular cone.

Example 128: Find the slant height of a cone with height 8 whose base has radius 6.

Solution:

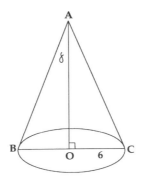

By the Pythagorean Theorem, we have $AC^2 = 6^2 + 8^2 = 100$. So the slant height is $AC = 10$.

Note: We can also use the Pythagorean triple 3, 4, 5 and observe that $6 = 2 \cdot 3$ and $8 = 2 \cdot 4$. So $AC = 2 \cdot 5 = 10$.

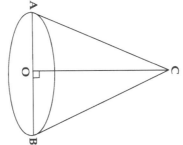

Example 129:

The figure above shows a cone with $OC = 4$ cm and $BC = 5$ cm. Find the perimeter of the lateral surface (the figure obtained when the cone is cut along a slant height) of the cone.

Solution:

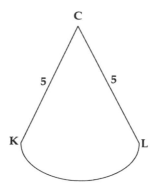

When we cut the cone along a slant height and open it up, we see that the perimeter consists of two slant heights together with the circumference of the base.

We use the Pythagorean triple 3, 4, 5 to see that the radius of the base is 3 cm. It follows that the circumference of the base is 6π cm.

Finally the perimeter is $5 + 5 + 6\pi = $ **10 + 6π cm.**

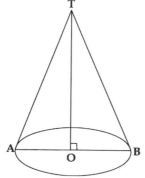

Example 130:

In the cone drawn above, $TB = 20$ cm and $TO = 16$ cm. Find the surface area of the cone.

Solution: By using a multiple of the Pythagorean triple 3, 4, 5 (or the Pythagorean Theorem) we see that the radius of the base circle is $r = OB = 4 \cdot 3 = 12$. So the area of the base is

$$B = \pi r^2 = \pi(12)^2 = 144\pi$$

and the lateral area is $L = \pi r a = \pi \cdot 12 \cdot 20 = 240\pi$.

So the surface area is $A = B + L = 144\pi + 240\pi = $ **384π cm².**

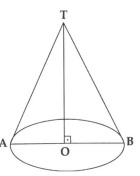

Example 131:

In the cone drawn above, $TO = 6$ cm and $OB = 8$ cm. Find the volume of the cone.

Solution: $V = \frac{1}{3}\pi r^2 h = \frac{1}{3}\pi \cdot 8^2 \cdot 6 = \mathbf{128\pi}$ **cm³.**

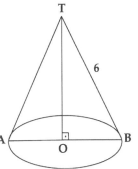

Example 132:

In the cone drawn above, $TB = 6$ cm and $OB = 2$ cm. Find the volume of the cone.

Solution: By the Pythagorean Theorem, $TB^2 = TO^2 + OB^2$. So $6^2 = h^2 + 2^2$. Therefore $h^2 = 36 - 4 = 32$, and so $h = \sqrt{32} = 4\sqrt{2}$.

Finally, $V = \frac{1}{3}\pi r^2 h = \frac{1}{3}\pi \cdot 2^2 \cdot 4\sqrt{2} = \frac{16\pi\sqrt{2}}{3}$ **cm³.**

Spheres

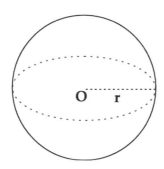

A **sphere** is a round solid figure surrounding a center point, every point on the surface being an equal distance from the center point. This distance is called the **radius** of the sphere.

The **surface area** of a sphere with radius r is $A = 4\pi r^2$.

The **volume** of a sphere with radius r is $V = \frac{4}{3}\pi r^3$.

Example 133: Find the surface area and volume of a sphere with radius 6 cm.

Solution: $A = 4\pi r^2 = 4\pi \cdot 6^2 = \mathbf{144\pi}$ **cm².**

$V = \frac{4}{3}\pi r^3 = \frac{4}{3}\pi \cdot 6^3 = \mathbf{288\pi}$ **cm³.**

Example 134: Find the surface area and volume of a sphere whose largest circle has an area of 25π cm².

Solution: The radius r of the sphere satisfies $\pi r^2 = 25\pi$. So $r^2 = 25$, and therefore $r = 5$ cm.

$A = 4\pi r^2 = 4\pi \cdot 5^2 = \mathbf{100\pi}$ **cm².**

$V = \frac{4}{3}\pi r^3 = \frac{4}{3}\pi \cdot 5^3 = \mathbf{\frac{500\pi}{3}}$ **cm³.**

Example 135: Find the surface area and volume of the solid generated by revolving a circle of radius 16 cm around its diameter 360°.

Solution:

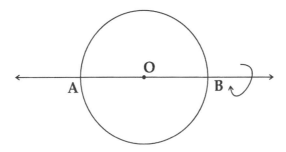

A sphere of radius 16 cm is generated.

$A = 4\pi r^2 = 4\pi \cdot 16^2 = \mathbf{1024\pi}$ **cm²**.

$V = \frac{4}{3}\pi r^3 = \frac{4}{3}\pi \cdot 16^3 = \frac{16{,}384\pi}{3}$ **cm³**.

Test – 1

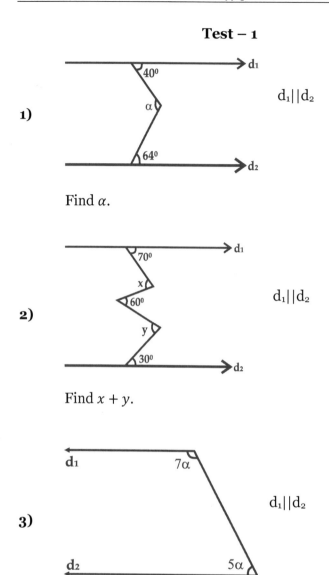

1)

$d_1 || d_2$

Find α.

2)

$d_1 || d_2$

Find $x + y$.

3)

$d_1 || d_2$

Find α.

4)

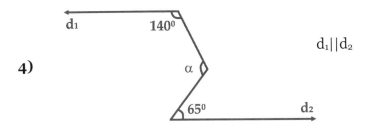

$d_1||d_2$

Find α.

5)

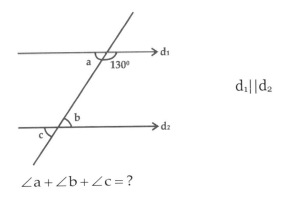

$d_1||d_2$

$\angle a + \angle b + \angle c = ?$

6)

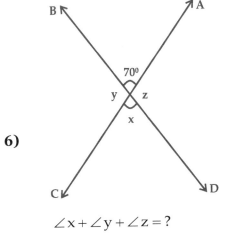

$\angle x + \angle y + \angle z = ?$

7) Two angles are supplementary and the measure of one is eight times the measure of the other. Find the measure of the smaller angle.

8) Two angles are supplementary and the measure of one is eleven times the measure of the other. Find the measure of the larger angle.

9) The difference between two supplementary angles is 20°. Find the measure of the smaller angle.

10) The ratio of one angle to another is 9:11. If the two angles are supplementary, find the angle measures.

11) The ratio of one angle to another is 3:2. If the two angles are complementary, find the angle measures.

12) The ratio of one angle to another is 1:5. If the two angles are complementary, find the difference between the two angle measures.

13) The difference between two complementary angles is 40°. Find the measure of the larger angle.

14) The difference between two complementary angles is 20°. Find the measure of the smaller angle.

15) The ratio of one angle to another is 5:4. If the two angles are complementary, find the difference between the two angle measures.

Test – 2

1)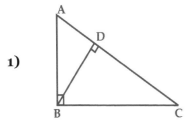

In the figure above, $m\angle B = m\angle D = 90°$, $m\angle C = 60°$, and $AB = \sqrt{3}$. Find the area of $\triangle ABC$.

2)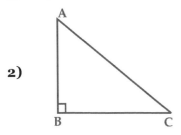

In the figure above, $m\angle B = 90°$, $AB = 7x + 2$, $BC = 7x + 4$, and $AC = 7x + 6$. Find the perimeter of $\triangle ABC$.

3)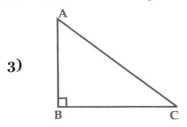

In the figure above, $m\angle B = 90°$, $m\angle C = 30°$, and $AC = 20$ cm. Find AB.

4)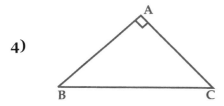

In the figure above, $m\angle A = 90°$, $m\angle B = 60°$, and $BC = 10$ cm. Find AC.

5)

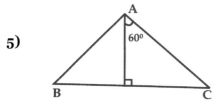

In the figure above, $m\angle A = 90°$, and $AB = 8$ cm. Find BC.

6)

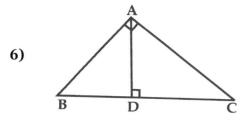

In the figure above, $m\angle A = m\angle ADC = 90°$, $m\angle C = 30°$, and $AB = 8$ cm. Find AD.

7)

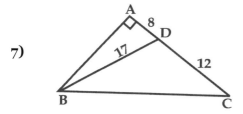

In the figure above, $m\angle A = 90°$, $AD = 8$, $DC = 12$, and $BD = 17$. Find BC.

8)

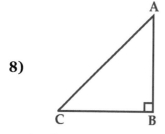

In the figure above, $m\angle B = 90°$, $AC = 61$ cm and $BC = 11$ cm. Find AB.

9)

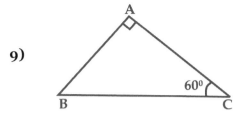

In the figure above, $m\angle A = 90°$, $m\angle C = 60°$, and $AC = 18$ cm. Find AB.

10)

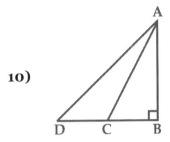

In the figure above, $m\angle B = 90°$, $AC = 6$ cm, $BC = 3$ cm, and $DC = 4$ cm. Find AD.

11)

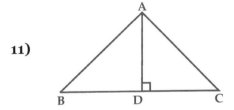

In the figure above, $m\angle D = 90°$, $DC = 3\sqrt{3}$, and $BD = 5\sqrt{3}$. Compute the ratio of the area of $\triangle ABD$ to the area of $\triangle ADC$.

12)

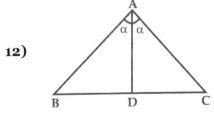

In the figure above, $m\angle BAD = m\angle DAC = \alpha$, $BD = 9$, and $DC = 7$. Compute the ratio of the area of $\triangle ABD$ to the area of $\triangle ADC$.

13)

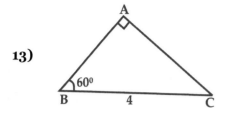

In the figure above, $m\angle A = 90°$, $m\angle B = 60°$, and $BC = 4$. Find the perimeter of $\triangle ABC$.

14)

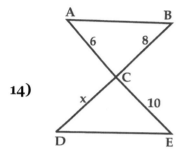

In the figure above, $AB \parallel DE$. Find x.

15)

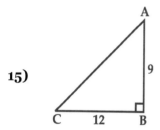

In the figure above, $AC = 2x$. Find the value of x.

Test – 3

1) The ratio of similarity of two similar triangles is 4:5. Find the ratio of the altitudes and ratio of the areas of the triangles.

2) Angle bisector BD of $\triangle ABC$ is drawn. $AB = 10$, $AD = 4$, and $DC = 3$. Find BC.

3) In $\triangle ABC$, $m\angle A = 80°$, $m\angle B = 40°$,and AD is an angle bisector. Find $m\angle ADC$.

4) Find the perimeter and area of an equilateral triangle with side of length 8.

5) Two sides of a triangle have lengths 5 and 8 cm. What is the largest possible integer value for the length of the third side?

6) In $\triangle ABC$, the median AD is drawn. Find the area of $\triangle ABC$ if the area of $\triangle ABD$ is 15 cm².

7) The ratio of the legs of a right triangle is 3:4. Find the perimeter of the triangle if the area of the triangle is 24cm².

8) In $\triangle ABC$, $m\angle A = 105°$, $m\angle B = 45°$, and $AC = 8$ cm. Find the area of $\triangle ABC$.

9) In $\triangle ABC$, $AB = AC = 10$ and $BC = 12$. Find the area of $\triangle ABC$.

10) In a 30°, 60°, 90° triangle, the side opposite to the 30° angle has length 6. Find the area and perimeter of the triangle.

11) Find the area and perimeter of the isosceles right triangle with hypotenuse $15\sqrt{2}$.

12) Find the area and perimeter of the right triangle with legs of length 10 and 24.

13) The ratio of the sides of a triangle is 1:2:3. Find the length of the smallest side if the perimeter of the triangle is 36 cm.

14) Find the area of an equilateral triangle with side length 9 cm.

15) Find the area of an equilateral triangle with perimeter 1 cm.

Test – 4

1) Find the number of sides of a polygon if the sum of the measures of its interior angles is 1800°

2) Find the number of sides of a polygon if the sum of the measures of its interior angles is 2160°.

3) Find the number of diagonals of a polygon with 8 sides.

4) Find the number of sides of a polygon with 54 diagonals.

5) Find the angle measure of an exterior angle of a regular polygon whose number of diagonals is three times its number of its sides.

6) Find the number of sides of a polygon whose sum of the measures of its interior angles is 1440°.

7) What is the angle measure of an interior angle of a regular 12-sided polygon?

8) Find the number of sides of a regular polygon whose interior angles measure 8 times each of its exterior angles.

9) Find the number of sides of a regular polygon if its exterior angle measures α and its interior angle measures $\frac{7\alpha}{2}$.

10) Find the number of sides of a regular polygon if the ratio of its interior to exterior angle measures is 7:2.

11) The ratio of the interior angles of a quadrilateral is 1: 2: 3: 4. Find the measure of the largest angle.

12) Find the number of sides of a regular polygon if each of its interior angles measures 150°.

13) Find the number of sides of a regular polygon if each of its interior angles measures 135°.

14) Find the number of sides of a regular polygon whose interior angles measure 60° more than the measure of one of its exterior angles.

15) Find the number of sides of a regular polygon if the difference between an interior and exterior angle measure is 150°.

Test – 5

1)   Find the perimeter of the square with side length 8 cm.

2)   Find the area of the square with diagonal of length $8\sqrt{2}$ cm.

3)   Find the area of the square with side length $\sqrt{7} + \sqrt{3}$.

4)   Find the length of the diagonal of the square with side length $\sqrt{3}$ cm.

5)

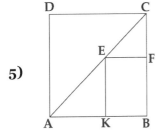

In square $ABCD$ above, $AB = 10$ cm. Find the perimeter of square $KBFE$.

6)

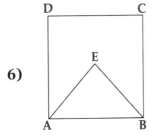

In the figure above, $ABCD$ is a square, ABE is an equilateral triangle, and $DC = 8$ cm. Find EB.

7)

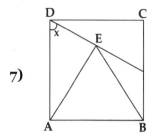

In the figure above, $ABCD$ is a square, and ABE is an equilateral triangle. Compute the value of x.

8)

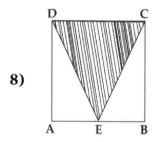

In the figure above, the area of triangle DEC is 18 cm². Find the perimeter of square $ABCD$.

9)

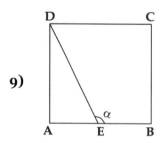

In the figure above, $ABCD$ is a square, $AE = 2$ cm, and $\alpha = 120°$. Find the area of $ABCD$.

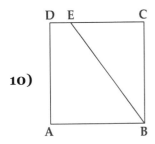

10)

In the figure above, $ABCD$ is a square, $EC = 6$, and $m\angle DEB = 120°$. Find DE.

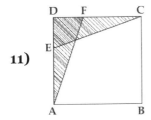

11)

In the figure above, $ABCD$ is a square, and we have $DF = FC = 4$ cm, $DE = EA = 4$ cm. Find the area of the shaded region.

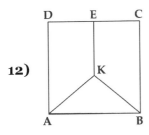

12)

In the figure above, $ABCD$ is a square, and $AK = KE = KB = 8$. Find AB.

13)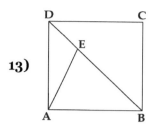

In the figure above, $ABCD$ is a square, $BC = 20$ cm, and $DE = 4\sqrt{2}$ cm. Find AE.

14)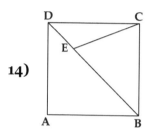

In the figure above, $ABCD$ is a square, BD is a diagonal of the square, $EC = 8\sqrt{3}$ cm, and $m\angle DCE = 15°$. Find AB.

15) Let H and S be a regular hexagon and a square, respectively, with equal perimeter. Find the ratio of the area of H to the area of S.

Test – 6

1) The perimeter of a rectangle is 80 cm. Find the area of the rectangle if the ratio of the side lengths is 5 : 3

2)

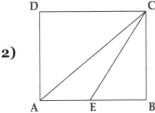

In the figure above, $ABCD$ is a rectangle, $AE = 3EB$, and the area of $\triangle CEB$ is 12 cm. Find the area of rectangle $ABCD$.

3)

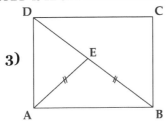

In the figure above, $ABCD$ is a rectangle, $AE = EB$, and $m\angle ADE = 54°$. Find $m\angle DEA$.

4)

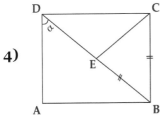

In the figure above, $ABCD$ is a rectangle, $BE = BC$, and $m\angle BCE = 70°$. Find $m\angle\alpha$.

5)

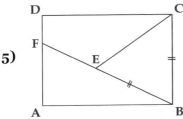

In the figure above, $ABCD$ is a rectangle, $BE = BC$, and $m\angle BEC = 64°$. Find $m\angle AFB$.

6) The length of a diagonal of a rectangle is 15 cm. Find the area of the rectangle if one of the side lengths is 12 cm.

7) The length of a diagonal of a rectangle is 26 cm. Find the perimeter of the rectangle if one of the side lengths is 10 cm.

8) One side of a rectangle is 18 cm long and forms a 30° angle with a diagonal. Find the length of the diagonal.

9) A rectangle with perimeter 28 cm has side lengths that are in the ratio of 3 to 4. Find the length of a diagonal of the rectangle.

10) The sides of a rectangle with perimeter 136 are in the ratio 5:12. Find the area of the rectangle.

11) The area of a rectangle with sides of length x and $x + 1$ is 12 cm². Find the perimeter of the rectangle.

12) The diagonal of a rectangle has length $\sqrt{65}$ cm. Find the area of the rectangle if its side lengths are x and $x + 3$.

13) The area of a rectangle is 60 cm² and its perimeter is 34 cm. Find the length of a diagonal of the rectangle.

14)

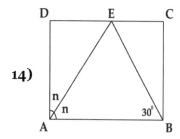

In the figure above, $DE = 10$, $m\angle ABE = 30°$, and AE is an angle bisector. Find AB.

15) The lengths of the sides of a rectangle are in the ratio 3:4. Find the area of the rectangle if its diagonal is 20 cm long.

Test-7

1) Find the area of a rhombus with side length 8 and altitude 6 cm.

2) Find the perimeter of a rhombus with altitude 4 cm an area 36cm².

3) Find the area of a rhombus with diagonals of length 10 and 16 cm.

4) Find the perimeter of a rhombus with diagonals of length 8 and 12 cm.

5) Find the sum of the lengths of the diagonals of a rhombus if its area is 40 cm² and one of the diagonals has length 8 cm.

6) The ratio of diagonal lengths of a rhombus is 3:4. Find the difference between the lengths of the diagonals if the area of the rhombus is 120 cm².

7) Find the perimeter of a rhombus with diagonals of length 5 and 12 cm.

8) Find the ratio of the areas of two rhombuses with equal altitudes and side length ratio of 3:7.

9) Find the perimeter of a rhombus with diagonal lengths $6\sqrt{3}$ and $8\sqrt{3}$ cm.

10) The area of a rhombus is 36 cm² and the lengths of its diagonals are in the ratio of 1:2. Find the perimeter of the rhombus.

11) Find the area of a rhombus with side length 20 cm and diagonal lengths in the ratio 3:4.

12) The side length of a rhombus is $2a + 2$, and the length of a diagonal is $4a$. Find the length of the other diagonal in terms of a.

13) Find the perimeter of a rhombus with area 24 cm² and such that the sum of the diagonal lengths is 20 cm.

14) The diagonals of a rhombus are 10 cm and 24 cm. Find its altitude.

15) Find the area of a rhombus with perimeter 40 cm and such that the sum of diagonal lengths is 28 cm.

Test – 8

1) Find the area of a parallelogram with base 12 cm and height 5 cm.

2) Find the perimeter of a parallelogram with sides of length 6 and 8 cm.

3) Find the sum of the squares of the diagonals of a parallelogram with sides of length 3 cm and 4 cm.

4)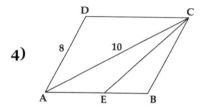

In the figure above $ABCD$ is a parallelogram, CE bisects $\angle BCA$, $AD = 8$, $AC = 10$, and $EB = 4$. Compute AB.

5)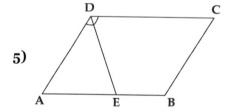

In the figure above $ABCD$ is a parallelogram, DE bisects $\angle ADC$, $EB = 6$ cm, and $BC = 16$ cm. Find the perimeter of $ABCD$.

6)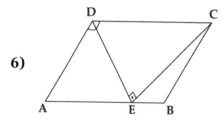

In the figure above $ABCD$ is a parallelogram, the area of $\triangle ADE$ is 12 cm², and the area of $\triangle BEC$ is 15 cm². Find the area of $ABCD$.

7)

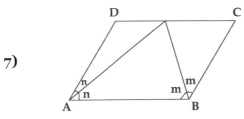

In the figure above *ABCD* is a parallelogram, and two angle bisectors are drawn. Given that *AB* = 16 cm, find the perimeter of *ABCD*.

8)

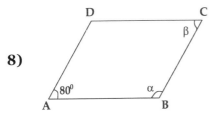

In the figure above *ABCD* is a parallelogram, $m\angle A = 80°$, $m\angle B = \alpha$, and $m\angle C = \beta$. Compute $\alpha - \beta$.

9)

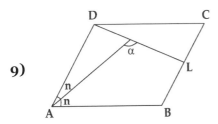

In the figure above, *ABCD* is a parallelogram, an angle bisector at $\angle A$ is drawn, $m\angle CDL = 25°$, and $m\angle C = 74°$. Find α.

10)

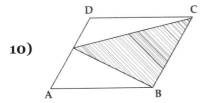

In the figure above *ABCD* is a parallelogram with area 42 cm². Compute the area of the shaded region.

11) The area of parallelogram $ABCD$ is 84 cm² and its height is 4 cm. Find the length of the corresponding base.

12)

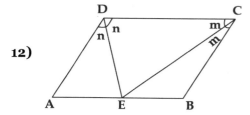

In the figure above $ABCD$ is a parallelogram, DE bisects $\angle D$, CE bisects $\angle C$, and the perimeter of $ABCD$ is 120 cm. Find AB.

13)

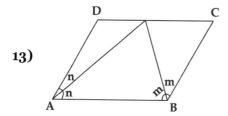

In the figure above $ABCD$ is a parallelogram, two angle bisectors are drawn, and $BC = 7$ cm. Find the perimeter of $ABCD$.

14) The ratio of the sides of a parallelogram is 3:8. What is the least possible integer value for its perimeter.

15)

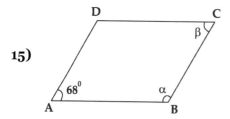

In the figure above $ABCD$ is a parallelogram, $m\angle A = 68°$, $m\angle B = \alpha$, and $m\angle C = \beta$. Evaluate $\alpha - \beta$.

Test – 9

1) Find the diameter of a circle with circumference 14π.

2) Find the circumference of a circle with radius 10 cm.

3) Find the ratio of the circumferences of two circles whose radii are in the ratio 3:4.

4) Find the circumference of a circle with diameter $2\sqrt{2}$ cm.

5) Find the length of the arc intercepted by a central angle of measure 90° in a circle with radius 5 cm.

6) The radius of a circle is 10 cm. Find the length of an arc intercepted by a central angle measuring 60°.

7) Find the circumference of a circle circumscribed about a square with side length $4\sqrt{2}$ cm.

8) Find the radius of a circle circumscribed about a square with side length 10 cm.

9)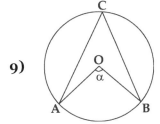

In the circle above with center O, $m\angle ACB = 40°$. Find $m\angle \alpha$.

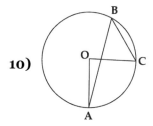

10)

In the circle above with center O, $m\angle AOC = 80°$. Find $m\angle ABC$.

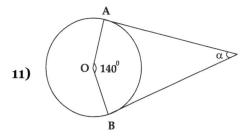

11)

In the circle above with center O, $m\angle BOA = 140°$. Find $m\angle\alpha$.

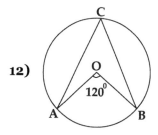

12)

In the circle above with center O, $m\angle ACB = x°$, $m\angle AOB = 120°$, and $m\widehat{AB} = y°$. Find $x + y$.

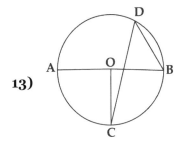

13)

In the circle above with center O, $m\angle AOC = 80°$. Find $m\angle COB + m\angle CDB$.

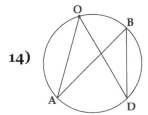

14)

In the figure above $m\angle AOD = 80°$. Find $m\angle ABD$.

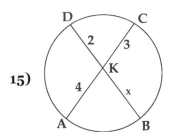

15)

In the figure above $AK = 4$, $KC = 3$, $DK = 2$, and $BK = x$. Find KB.

Test – 10

1) Find the diameter of circle with area 49π cm².

2) Find the radius of a circle with area π cm²·

3) Find the radius of a circle with area 81π cm².

4) Find the ratio of the areas of the circles with radii in the ratio 3:5.

5) Find the sum of the radii of circles with areas 64π cm² and 100π cm².

6)

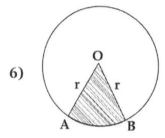

In the figure above $r = 10$ cm, and $m\angle AOB = 80°$. Find the area of the shaded sector.

7)

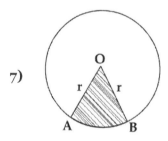

In the figure above $r = 12$ cm, and $m\angle AOB = 30°$. Find the area of the shaded sector.

8)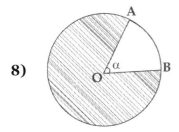

The circle with center O shown above has a diameter of 12 cm. If the area of the shaded region is 33π, find α.

9)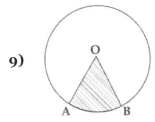

In the circle drawn above with center O, $\overset{\frown}{AB} = 60°$ and $OA = 8$ cm. Find the area of the shaded region.

10)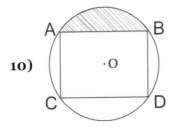

In the circle drawn above with center O, $AB = 8$ cm and $ABDC$ is a square. Find the area of the shaded region.

11) Find sum of the radii of two circles with areas 25π cm² and 81π cm².

12) Find 20% of the radius of a circle with area 400π.

13) Find the area of a circle with radius $\sqrt{3}$ cm.

14) By what percent will the area of circle decrease if its radius is decreased by 20%?

15) By what percent will the area of a circle increase if its diameter is increased by 40%?

Test – 11

1)

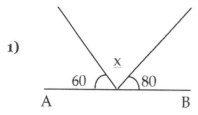

Find x in the figure above.

2)

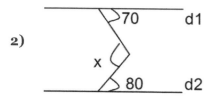

In the figure above, $d_1 \parallel d_2$. Find x.

3) What is the smallest angle in a triangle whose angles are in the ratio 1:3:5?

4)

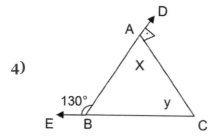

In the triangle above, $m\angle EBA = 130°$, and $m\angle DAC = 90°$. Find $x + y$.

5)

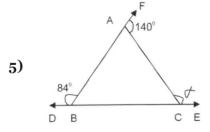

In the triangle above, $m\angle FAC = 140°$, and $m\angle ABD = 84°$. Find α.

6)

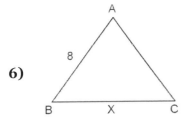

In the triangle drawn above, $AB = 8$ and $x = 5$. What is the greatest possible integer value for AC?

7)

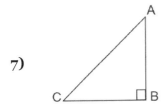

In the triangle drawn above, $AC = 15$, $BC = 12$, $AB = 3x$, and $m\angle B = 90°$. Find x.

8)

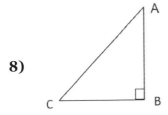

In the triangle drawn above, $AC = 20$, $AB = 12$, $BC = 4x$, and $m\angle B = 90°$. Find x.

9)

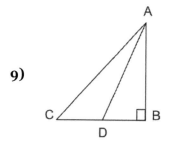

In the triangle drawn above, $AB = 4$, $AD = 8$, $DC = 2\sqrt{3}$ and $m\angle B = 90°$. Find AC.

10)

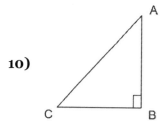

In the triangle drawn above, $AC = 12$, $AB = 6$, and $m\angle B = 90°$. Find the perimeter of $\triangle ABC$.

11)

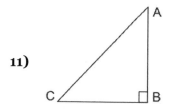

In the triangle drawn above, $AC = 10$ cm, $m\angle C = 30°$, and $m\angle B = 90°$. Find the perimeter of $\triangle ABC$.

12)

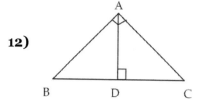

In the triangle drawn above, $m\angle A = m\angle ADC = 90°$, $BD = 4$, and $DC = 8$. Find AD.

13)

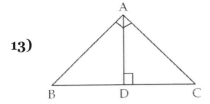

In the triangle drawn above, $m\angle A = 90°$, $BD = 3$, and $DC = 7$. Find AD.

14)

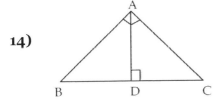

In the triangle drawn above, $m\angle A = m\angle ADC = 90°$, $AD = 6$, and $BD = 3$. Find DC.

15)

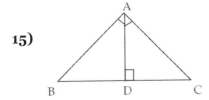

In the triangle drawn above, $m\angle A = m\angle ADC = 90°$, $BD = 4$, and $AD = \sqrt{12}$. Find DC.

Test – 12

1)

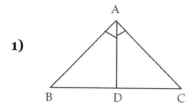

In the triangle drawn above, $m\angle A = m\angle D = 90°$, $BD = 6$, and $DC = 8$. Find the perimeter of $\triangle ABC$.

2) What is the area of an equilateral triangle with edge length 8.

3)

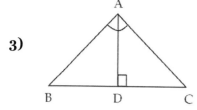

In the triangle drawn above, AD bisects $\angle A$, $AB = 8$, $AC = 10$, and $BD = 5$. Find DC.

4)

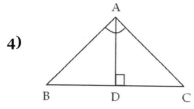

In the triangle drawn above, AD bisects $\angle A$, $AB = 8$, $BC = 12$, and $AC = 6$. Find DC.

5)

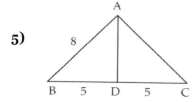

In the triangle drawn above, $m\angle A = 90°$, $AB = 8$, $BD = 5$, and $DC = 5$. Find AD.

6)

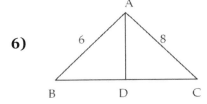

In the triangle drawn above, AD bisects $\angle A$, $AB = 6$ cm, $AC = 8$ cm, and the area of $\triangle ABD$ is 12 cm². Find the area of $\triangle ADC$.

7)

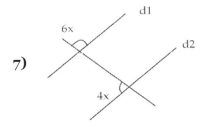

In the figure above, $d_1 \parallel d_2$. Find x.

8)

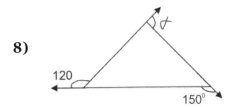

Find α in the figure above.

9)

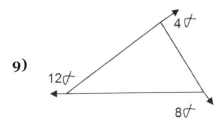

Find α in the figure above.

10)

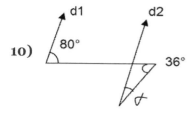

In the figure above, $d_1 \parallel d_2$. Find α.

11)

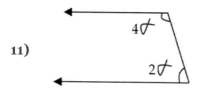

In the figure above, $d_1 \parallel d_2$. Find α.

12)

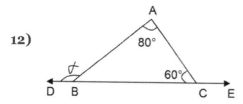

Find α in the figure above.

13)

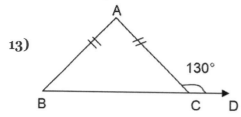

In the figure above, $AB = AC$ and $m\angle ACD = 130°$. Find $m\angle A$.

14)

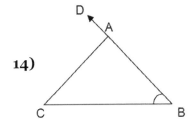

In the figure above, $m\angle DAC = 120°$, $m\angle B = 6\alpha$ and $m\angle C = 4\alpha$. Find α.

15)

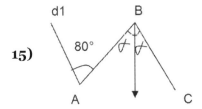

In the figure above, $d_1 \parallel BC$. Find α.

Test – 13

1)

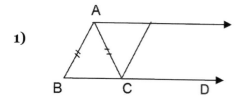

In the figure above, $m\angle BAC = 70°$, and $AB = AC$. Find $m\angle ACD$.

2)

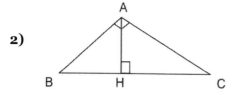

In the figure above, $m\angle HAC = 30°$, $m\angle A = m\angle AHC = 90°$, and $AC = 8$. Find BH.

3)

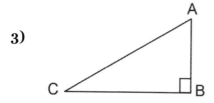

In the figure above, $m\angle B = 90°$, $m\angle A = 45°$, $AB = 4$ cm. Find AC.

4)

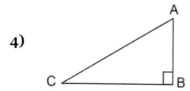

In the figure above, $m\angle B = 90°$, $m\angle A = 45°$, $AC = 10$ cm. Find AB.

5)

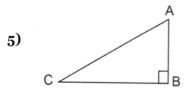

In the figure above, $m\angle B = 90°$, $m\angle A = 45°$, $AC = 5\sqrt{2}$ cm. Find the area of $\triangle ABC$.

6)

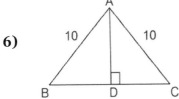

In the figure above, $m\angle D = 90°$, $AB = AC = 10$ cm, and $BC = 12$ cm. Find AD.

7)

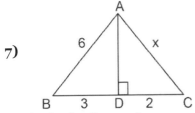

Find x in the figure above.

8)

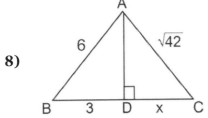

Find x in the figure above.

9)

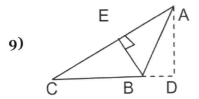

In the figure above, $CE = AE = 4$ cm, and $BE = 3$ cm. Find AB.

10)

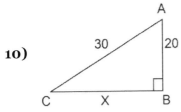

Find x in the figure above.

11)

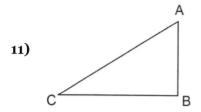

In the figure above, $m\angle B = 90°$, $m\angle C = 30°$, $AB = 6$ cm. Find BC.

12)

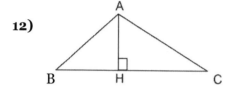

In the figure above, $m\angle A = 90°$, $m\angle C = 25°$. Find $m\angle ABH$.

13)

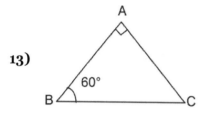

In the figure above, $BC = 10$ cm. Find AC.

14)

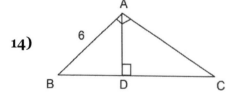

In the figure above, $m\angle C = 30°$. Find BC.

15)

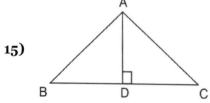

In the figure above, $m\angle A = 90°$, $m\angle DAC = 30°$, and $AC = 4\sqrt{3}$ cm. Find BC.

Test – 14

1)

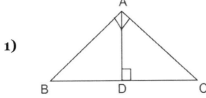

In the figure above, $BD = 1$ cm and $DC = 3$ cm. Find AD.

2)

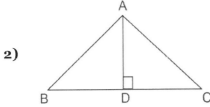

In the figure above, $m\angle A = 90°$, $BD = 3$ and $DC = 6$. Find the perimeter of $\triangle ABC$.

3)

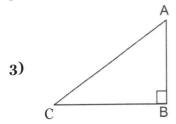

In the figure above, $AB = 4$ cm and $AC = 6$ cm. Find BC.

4)

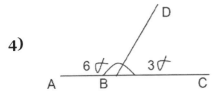

Find α in the figure above.

5)

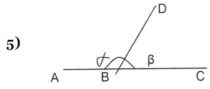

In the figure above, $\alpha - \beta = 30$. Find α.

6)

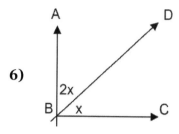

In the figure above, $m\angle B = 90°$. Find x.

7)

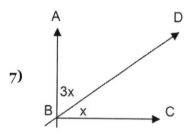

In the figure above, $m\angle B = 90°$. Find x.

8)

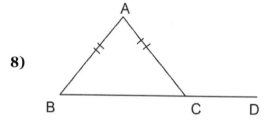

In the figure above, $m\angle ACD = 130°$ and $AB = AC$. Find $m\angle A$.

9) The exterior angles of a triangle are in the ratio 3:5:7. Find the largest of these three angles.

10)

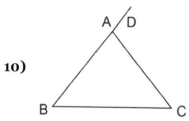

In the figure above, $\frac{m\angle B}{m\angle C} = \frac{2}{3}$ and $m\angle DAC = 120°$. Compute $m\angle C - m\angle B$.

11)

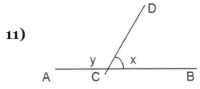

In the figure above, $m\angle y - m\angle x = 20°$. Find $m\angle y$.

12)

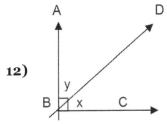

In the figure above, $m\angle y - m\angle x = 10°$. Find $m\angle y$.

13)

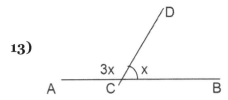

Find x in the figure above.

14)

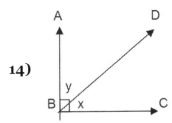

In the figure above, $\frac{x}{y} = \frac{1}{5}$. Find x.

15)

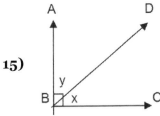

In the figure above, $\frac{x}{y} = \frac{1}{8}$. Find y.

Test – 15

1) Find the area of an equilateral triangle with a side of length 5.

2) Find the area of an equilateral triangle with a side of length 12.

3)
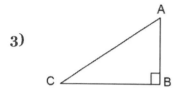

In the figure above, $AB = 8$ and $BC = 12$. Find the area of $\triangle ABC$.

4)
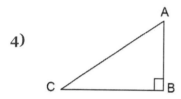

In the figure above, $AB = 14$ cm and $BC = 12$ cm. Find the area of $\triangle ABC$.

5)
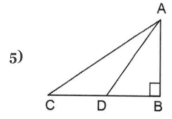

In the figure above, $AC = 17$ cm, $CD = 2$ cm, and $AB = 15$ cm. Find the area of $\triangle ADC$.

6)
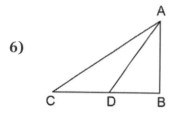

In the figure above, $AB = 10$ cm and $CD = 4$ cm. Find the area of $\triangle ADC$.

7)

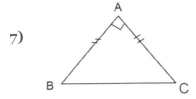

In the figure above, $BC = 6\sqrt{2}$ cm. Find the area of $\triangle ABC$.

8)

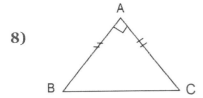

In the figure above, $BC = 2$ cm. Find the area of $\triangle ABC$.

9)

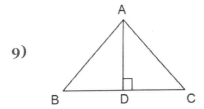

In the figure above, $BD = 8$ and $DC = 10$. Find the ratio of the area of $\triangle ABC$ to the area of $\triangle ADC$.

10) Find the area of an equilateral triangle with a side of length 9 .

11)

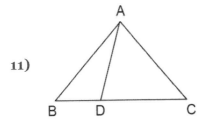

In the figure above, $BD = 2$ cm, $DC = 4$ cm, and the area of $\triangle ABD$ is 6 cm². Find the area of $\triangle ABC$.

12)

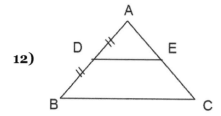

In the figure above, D is the midpoint of AB, E is the midpoint of AC, and the area of $\triangle ADE$ is 8 cm². Find the area of $\triangle ABC$.

13)

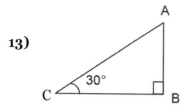

In the figure above, $AC = 1$ cm. Find the area of $\triangle ABC$.

14)

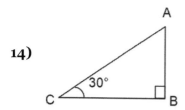

In the figure above, $AC = 4\sqrt{3}$ cm. Find the area of $\triangle ABC$.

15)

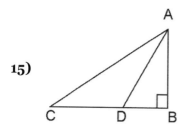

In the figure above, $AC = 20$ cm, $CD = 6$ cm, and $AB = 12$ cm. Find the area of $\triangle ADB$.

Test – 16

1) A polygon has 6 sides. What is the sum of the measures of the interior angles?

2) A polygon has 10 sides. What is the sum of the measures of the interior angles?

3) How many diagonals does an 8 sided polygon have?

4) How many diagonals does an 10 sided polygon have?

5) The measure of an exterior angle of a regular polygon is 30°. How many sides does the polygon have?

6) The sum of the measures of the interior angles of a polygon is 2160°. How many sides does the polygon have?

7) The sum of the measures of the interior angles of a regular polygon with edge length 4 is 1440°. Find the perimeter.

8) What is the degree measure of an exterior angle of a regular octagon?

9) The sum of the interior angles of a regular polygon with side length 10 cm is 540. Find the perimeter.

10) The positive difference between the measures of an internal and external angle of a polygon is 140°. Find the number of sides of the polygon.

11) Find the area of a square with a diagonal of length $6\sqrt{2}$.

12) Find the perimeter of a square with a diagonal of length $\sqrt{7}$.

13) Find the area of a square with perimeter 7 cm.

14) A diagonal of a square has length $\sqrt{6}$. Find the ratio of the perimeter of the square to its area.

15) The area of a square is 6 cm². What is the perimeter of the square?

Test – 17

1)

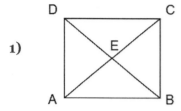

In the figure above, $ABCD$ is a square and $CE = 6$ cm. Find the area of $ABCD$.

2)

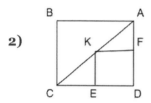

In the figure above, $ABCD$ and $KEDF$ are squares, $AB = 10$ cm, and K is the midpoint of AC. Find the area of $KEDF$.

3)

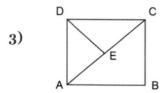

In the figure above, $ABCD$ is a square, $CE = 6$ cm, and $AE = 16$ cm. Find DE.

4)

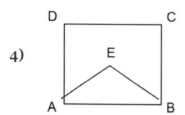

In the figure above, $ABCD$ is a square, $\triangle ABE$ is equilateral, and $AD = 8$ cm. Find EB.

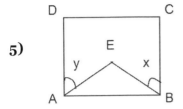

5)

In the figure above $ABCD$ is a square and $\triangle ABE$ is equilateral. Compute $x + y$.

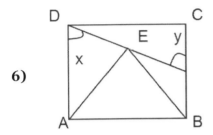

6)

In the figure above $ABCD$ is a square and $\triangle ABE$ is equilateral. Compute $x + y$.

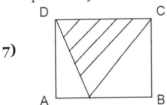

7)

In the figure above $ABCD$ is a square and the area of the shaded triangle is 12 cm². Find the area of $ABCD$.

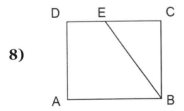

8)

In the figure above $ABCD$ is a square, $m\angle DEB = 120°$, and $EC = 4$. Find DE.

9)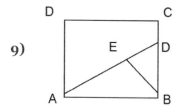

In the figure above $ABCD$ is a square, $m\angle AEB = 90°$, $AE = 6$, and $ED = 3$. Find the area of the square.

10)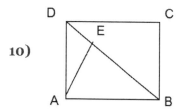

In the figure above $ABCD$ is a square, $BC = 12$ cm and $DE = 2\sqrt{2}$ cm. Find AE.

11)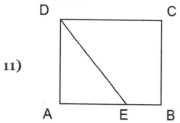

In the figure above $ABCD$ is a square, $m\angle DEB = 120°$, and $AE = 2$. Find the area of $ABCD$.

12) The perimeter of a square is 1 cm. Find the length of a diagonal of the square.

13) The length of a diagonal of a square is 2 cm. Find the perimeter of the square.

14) Two squares have diagonals with lengths in the ratio 2:5. Find the ratio of the perimeter of the smaller square to the perimeter of the larger square.

15) A square has a perimeter of $4\sqrt{3}$ cm. Find the length of a diagonal of the square.

Test – 18

1)   A rectangle has edges with lengths of 6 cm and 4 cm. Find the length of a diagonal of the rectangle.

2)   The lengths of two sides of a rectangle are in the ratio 2:3. Find the area of the rectangle if its perimeter is 50 cm.

3)   The lengths of two sides of a rectangle are in the ratio 3:4. Find the length of a diagonal of the rectangle if its area is 144.

4)   One of the edges of a rectangle has length 3 cm and a diagonal of the rectangle has length 7 cm. Find the rectangle's area.

5)   The perimeter of a rectangle is 60 cm. If the lengths of two sides are in the ratio 1 to 4, find the area of the rectangle.

6)

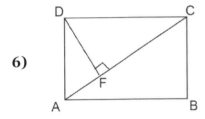

In the figure above $ABCD$ is a rectangle, $AF = 3$ and, $FC = 6$. Find the area of $ABCD$.

7)

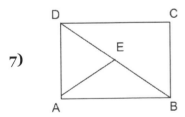

In the figure above $ABCD$ is a rectangle, $m\angle ADB = 54°$, and $AE = EB$. Compute $m\angle AED$.

8)

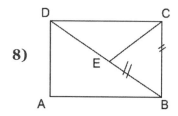

In the figure above $ABCD$ is a rectangle, $m\angle BEC = 60°$, and $EC = 6$ cm. Find the area of $\triangle BEC$.

9)

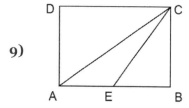

In the figure above $ABCD$ is a rectangle, $AE = 3EB$, and the area of $\triangle CEB$ is 6 cm². Find the area of $ABCD$.

10)

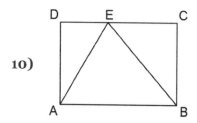

In the figure above $ABCD$ is a rectangle, $m\angle ABE = 30°$, AE bisects $\angle A$, and $BE = 8$ cm. Find AB.

11)

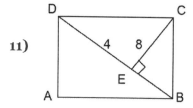

In the figure above $ABCD$ is a rectangle, $m\angle BEC = 90°$, $CE = 8$ cm, and $DE = 4$ cm. Find the area of $ABCD$.

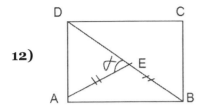

12)

In the figure above *ABCD* is a rectangle, $m\angle ADB = 70°$, and $AE = EB$. Find $m\angle AED$.

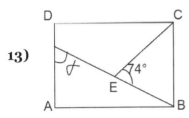

13)

In the figure above *ABCD* is a rectangle, $m\angle CEB = 74°$, and $CB = BE$. Find $m\angle\alpha$.

14) Two edges of a rectangle have lengths that are in the ratio 3:4. If the perimeter of the rectangle is 28 cm, find the length of a diagonal of the rectangle.

15) A rectangle with a side of length 10 cm has a perimeter of 60 cm. Find the length of the longest side of the rectangle.

Test – 19

1)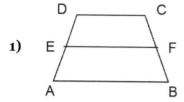

In the figure above $ABCD$ is a trapezoid, EF is the median, $DC = 8$, and $AB = 16$. Find EF.

2)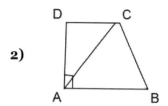

In the figure above $ABCD$ is a right trapezoid with $\angle A$, $\angle D$ and $\angle ACB$ are right angles. If $DC = 6$ and $AB = 10$, find AD.

3)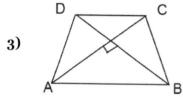

In the figure above $ABCD$ is an isosceles trapezoid, $DC = 10$ cm and $AB = 16$ cm. Find the area of $ABCD$.

4)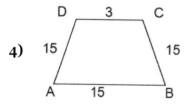

In the figure above $ABCD$ is a trapezoid, $DC = 3$ cm and $AB = AD = BC = 15$ cm. Find the area of $ABCD$.

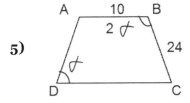

5)

In the figure above $ABCD$ is an isosceles trapezoid. Find DC.

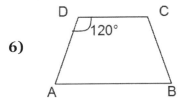

6)

In the figure above $ABCD$ is a trapezoid with $m\angle D = 120°$, and $AD = BC = DC = 8$ cm. Find the area of $ABCD$.

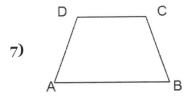

7)

In the figure above $ABCD$ is a trapezoid with $m\angle A = \alpha$, $m\angle C = 2\alpha$, $AD = 10$ cm, $DC = 9$ cm, and $AB = 18$ cm. Find BC.

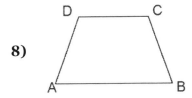

8)

In the figure above $ABCD$ is a trapezoid, $m\angle A = 45°$, $m\angle B = 30°$, $DC = 4$ cm, and $BC = 6$ cm. Find AB.

9)

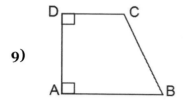

In the figure above $ABCD$ is a right trapezoid, $AB = BC = 15$ cm, and $DC = 6$ cm. Find the area of $ABCD$.

10)

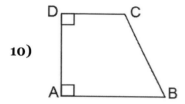

In the figure above $ABCD$ is a right trapezoid, $DC = BC = 10$ cm, and $m\angle C = 120°$. Find the area of $ABCD$.

11)

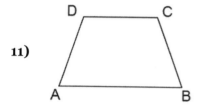

In the figure above $ABCD$ is a trapezoid, $DC = 5$, $BC = 8$, $AB = 13$, and $m\angle A = 50°$. Find $m\angle C$.

12)

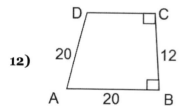

In the figure above $ABCD$ is a right trapezoid. Find DC.

13)

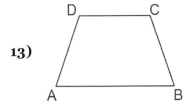

In the figure above $ABCD$ is a trapezoid, $m\angle A = 30°$, $m\angle B = 60°$, $DC = 6\sqrt{3}$, and $BC = 8\sqrt{3}$. Find AB.

14)

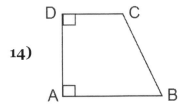

In the figure above $ABCD$ is a right trapezoid, $DC = 8$, $AD = 16$, and $AB = BC$. Find the area of $ABCD$.

15) Find the area of an equilateral triangle with side length 16 cm.

Test – 20

1) Find the area of a rhombus with diagonal lengths 8 and 12.

2) A rhombus has area 120 cm2 and a diagonal of length 12 cm. Find the length of the other diagonal.

3) The diagonals of a rhombus are in the ratio 1:3. If the area of the rhombus is 300 cm2, find the shorter diagonal length.

4) Find the perimeter of a rhombus with diagonal lengths 12 and 20.

5) The lengths of the diagonals of a rhombus are 10 cm and 24 cm. Find the perimeter of the rhombus.

6) One side of a rhombus has length 15 cm. The longer diagonal has length 24 cm. Find the length of the shorter diagonal.

7) A rhombus has diagonal lengths $8\sqrt{3}$ cm and $12\sqrt{3}$ cm. Find the side length of the rhombus.

8) A rhombus has side length 25 cm and diagonal lengths in the ratio 3:4. Find the area of the rhombus.

9) The perimeter of a rhombus is 48 cm and the lengths of the diagonals add up to 26 cm. Find the area of the rhombus.

10) The area of a rhombus is 40 cm2 and the sum of the diagonal lengths of the rhombus is 18 cm. Find the perimeter.

11) The area of a rhombus is 12 cm2 and the sum of the diagonal lengths is 10 cm. Find the side length of the rhombus.

12) The lengths of the diagonals of a rhombus are 30 cm and 16 cm. Find the length of the side of the rhombus.

13) The product of the lengths of the diagonals of a rhombus is $8\sqrt{3}$. Find the area of the rhombus.

14) Find the perimeter of a rhombus with diagonal lengths 18 cm and 24 cm.

15) The diagonal lengths of a rhombus are 10 cm and 24 cm. Find the height of the rhombus.

Test – 21

1)

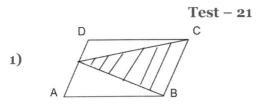

In the figure above *ABCD* is a parallelogram, and the area of the shaded triangle is 12 cm². Find the area of *ABCD*.

2)

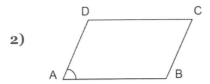

In the figure above *ABCD* is a parallelogram, $m\angle A = 70°$, and $m\angle D = 2x$. Find x.

3)

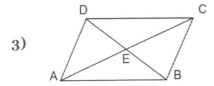

In the figure above *ABCD* is a parallelogram, and the area of $\triangle AEB$ is $4\sqrt{2}$ cm². Find the area of *ABCD*.

4)

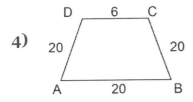

Find the height of trapezoid *ABCD* draw above.

5)

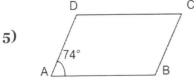

In the figure above *ABCD* is a parallelogram, and $m\angle A = 74°$. Compute $m\angle B - m\angle C$.

6)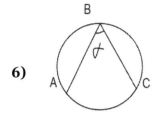

In the circle drawn above, $m\widehat{AC} = 74°$. Find α.

7)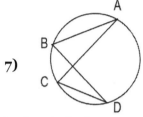

In the circle drawn above, $m\angle B = 60°$, and $m\angle C = 2x$. Find x.

8)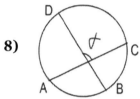

In the circle drawn above, $m\widehat{AB} = 60°$, and $m\widehat{DC} = 100°$. Find α.

9)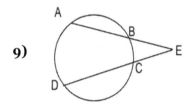

In the circle drawn above, $m\widehat{AD} = 80°$, and $m\widehat{BC} = 30°$. Find $m\angle E$.

10)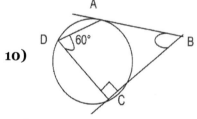

In the circle drawn above, $m\angle D = 60°$. Find $m\angle B$.

11)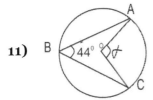

In the circle drawn above with center o, $m\angle B = 44°$. Find α.

12)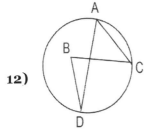

In the circle drawn above with center B, $m\angle B = 84°$. Find $m\angle A$.

13)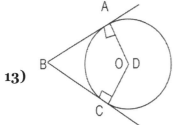

In the circle drawn above with center D, $m\angle D = 132°$. Find $m\angle B$.

14)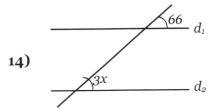

In the figure above, $d_1 \parallel d_2$. Find x.

15)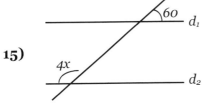

In the figure above, $d_1 \parallel d_2$. Find x.

Test - 22

1)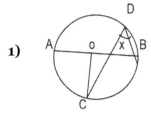

In the circle drawn above with center O, $m\angle AOC = 80°$. Find x.

2)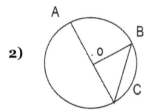

In the circle above with center O, $m\angle OBC = 48°$. Find $m\angle AOB$.

3)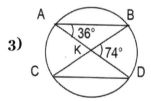

Find $m\angle KDC$ in the circle above.

4)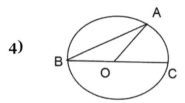

In the circle above with center O, $m\angle ABC = 30°$ and the radius of he circle is $r = 4$ cm. Find the length of $\widehat{AC}$.

5)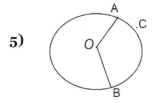

In the circle above with center O, $m\angle AOB = 120°$ and $r = 8$ cm. Find the length of $\overset{\frown}{ACB}$.

6)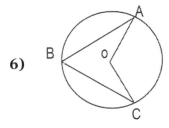

In the circle above with center O, $m\angle AOC = 110°$, $m\overset{\frown}{AC} = x$, and $m\angle B = y$. Find $x + y$.

7)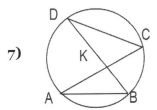

In the circle above, $m\angle D = 40°$ and $m\angle AKD = 74°$. Find $m\angle B$.

8)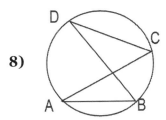

In the circle above, $m\angle B = 70°$. Find $m\angle C$.

9)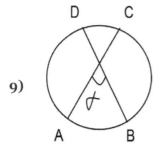

In the circle above, $m\widehat{DC} = 72°$ and $m\widehat{AB} = 64°$. Find α.

10)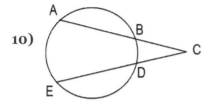

In the circle above $m\widehat{AE} = 110°$ and $m\widehat{BD} = 34°$. Find $m\angle C$.

11)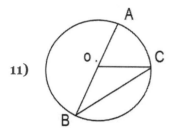

In the circle above $m\angle AOC = 64°$. Find $m\angle C$.

12)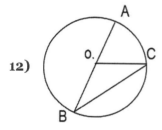

In the circle above $m\angle B = 33°$. Find $m\angle AOC$.

13)

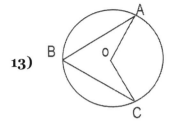

In the circle above with center O, $m\angle ABC = 64°$. Find $m\angle AOC$.

14)

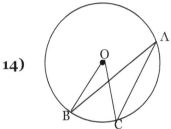

In the circle above with center O, $m\angle BOC = 52°$. Find $m\angle A$.

15)

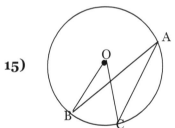

In the circle above with center O, $\overset{\frown}{BC} = 74°$. Compute $m\angle BOC + m\angle BAC$.

Test – 23

1) Find the radius of a circle with circumference 30π cm.

2) Find the diameter of a circle with area 169π cm².

3) A circle has a diameter of 10 cm. Find the area of a sector of the circle intercepted by a 90° central angle.

4)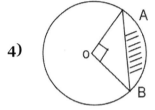

In the circle above with center O, $m\angle AOB = 90°$ and $OA = OB = 10$ cm. Find the area of the shaded region.

5)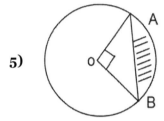

In the circle above with center O, $m\angle AOB = 90°$ and $AB = 8\sqrt{2}$ cm. Find the area of the shaded region.

6)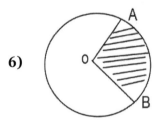

In the circle above with center O, $m\angle AOB = 72°$ and $OA = 6$ cm. Find the area of the shaded region.

7) Find the diameter of a circle with area 2π cm².

8) The radius of a circle is 8 cm. Find the length of the part of the circle intercepted by a 90° angle.

9)

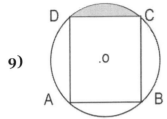

The figure above shows square $ABCD$ with its circumsribed circle with center O. If $AB = 6$ cm, find the shaded area.

10) The length of the radius of a circle is 14 cm. Find the area.

11) Find the diameter of a circle with area π cm².

12) Find the diameter of a circle with area 400π cm².

13)

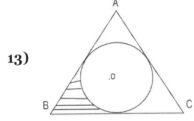

The figure above shows equilateral triangle ABC with its inscribed circle with center O. If $BC = 40$, find the shaded area.

14)

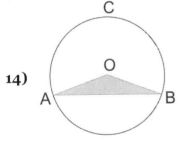

In the circle above with center O, $m\widehat{ACB} = 300°$ and $OA = 8$ cm. Find the area of the shaded region.

15) Two circles have radii in the ratio 4:7. Find the ratio of the areas of the circles.

Test - 24

1)

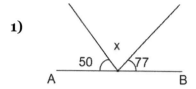

Find x in the figure above.

2)

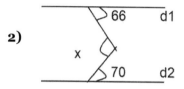

In the figure above $d_1 \parallel d_2$. Find x.

3)

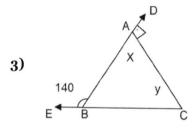

Find $x + y$ in the figure above.

4)

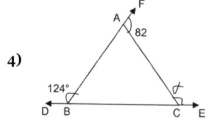

Find α in the figure above.

5)

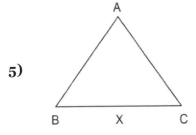

In the figure above, $AB = 8$ and $AC = 11$. Find the largest possible integer value of x.

6)

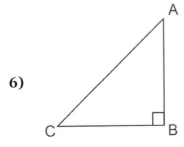

In the figure above, $AC = 15$, $AB = 3x$, and $BC = 12$. Find x.

7)

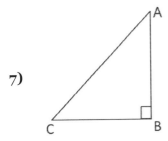

In the figure above, $AC = 20$, $AB = 16$, and $BC = 4x$. Find x.

8)

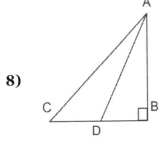

In the figure above, $AB = 5$, $AD = 13$, and $DC = 3$. Find AC.

9)

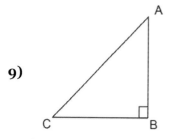

In the figure above, $AB = BC = 11$. Find the perimeter of $\triangle ABC$.

10)

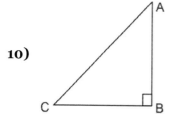

In the figure above, $AC = 12$ and $m\angle A = 60°$. Find the perimeter of $\triangle ABC$.

11)

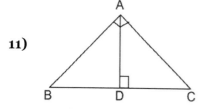

In the figure above, $BD = 6$ cm and $DC = 10$ cm. Find AD.

12)

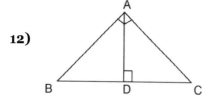

In the figure above, $AD = 8$ cm and $BD = 4$ cm. Find BC.

13)

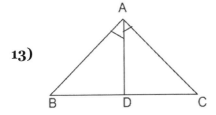

In the figure above, $m\angle A = m\angle ADC = 90°$, $BD = 6$, and $DC = 8$. Find the perimeter of $\triangle ABC$.

14)

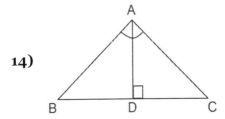

In the figure above, $AB = 6$ cm, $AC = 8$ cm, $BD = 3$ cm, and AD bisects $\angle A$. Find BC.

15)

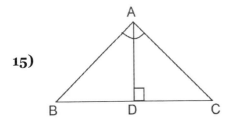

In the figure above, $AB = 10$, $AC = 8$, $BC = 14$, and AD bisects $\angle A$. Find DC.

Test – 25

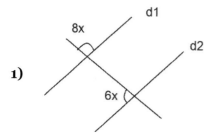

1)

In the figure above $d_1 \parallel d_2$. Find x.

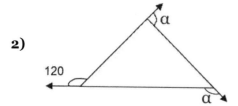

2)

Find α in the figure above.

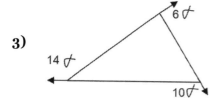

3)

Find α in the figure above.

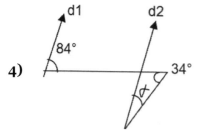

4)

In the figure above $d_1 \parallel d_2$. Find α.

5)

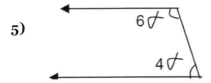

In the figure above $d_1 \parallel d_2$. Find 2α.

6)

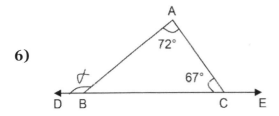

Find α in the figure above.

7)

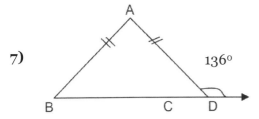

In the figure above, $AB = AC$. Find $m\angle A$.

8)

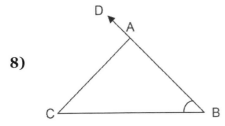

In the figure above, $m\angle B = 8\alpha$, $m\angle C = 6\alpha$, and $m\angle DAC = 120°$. Find α.

9)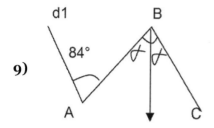

In the figure above $d_1 \parallel d_2$. Find α.

10)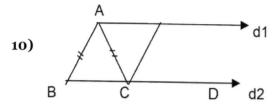

In the figure above, $AB = AC$ and $m\angle BAC = 64°$. Find $m\angle ACD$.

11)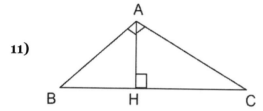

In the figure above, $m\angle HAC = 30°$ and $AC = 6$ cm. Find BC.

12)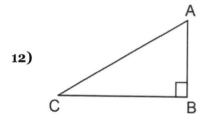

In the figure above, $m\angle A = 45°$ and $AB = 2$ cm. Find AC.

13)

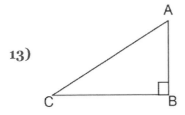

In the figure above, $m\angle A = 45°$ and $AC = 12$ cm. Find AB.

14)

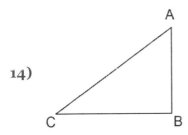

In the figure above, $m\angle B = 90°$, $m\angle A = 45°$ and $AC = 8\sqrt{2}$ cm. Find the perimeter of $\triangle ABC$.

15)

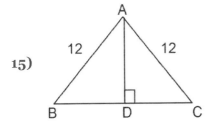

In the figure above, $AB = AC = 12$ cm and $BC = 14$ cm. Find AD.

Test – 26

1)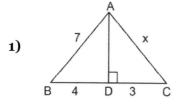

Find x in the figure above.

2)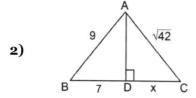

Find x in the figure above.

3)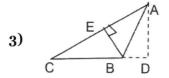

In the figure above, $CE = EA = BE = 6$ cm. Find AD.

4)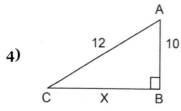

Find x in the figure above.

5)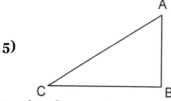

In the figure above, $m\angle B = 90°$, $m\angle C = 30°$, and $AB = 4$ cm. Compute $AC + BC$.

6)

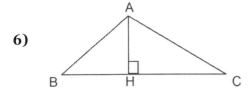

In the figure above, $m\angle A = 90°$, $m\angle AHC = 90°$, and $m\angle C = 24°$. Find $m\angle ABH$.

7)

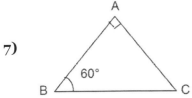

In the figure above, $BC = 24$. Compute $AC + AB$.

8)

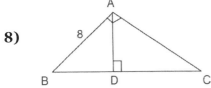

In the figure above, $m\angle C = 30°$. Compute $AC + BC$.

9)

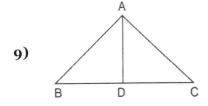

In the figure above, $m\angle A = 90°$, $m\angle ADC = 90°$, $BD = 4$ cm, and $DC = 7$ cm. Find the area of $\triangle ABC$.

10)

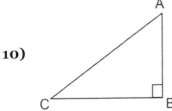

In the figure above, $AB = 3$ cm and $AC = 4$ cm. Find the perimeter of $\triangle ABC$.

11) Write an equation of the line with slope 4 and passing through the points (3,6) in point-slope form.

12) Write the equation of the line passing through the points (7,2) and (9,4) in slope-intercept form.

13)

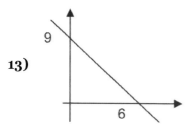

Write an equation of the line whose graph is shown above in slope-intercept form.

14) Write an equation of the line passing through the point (0,0) and parallel to the line $6x + 5y + 10 = 0$.

15) Write an equation of the line passing through the point $(-5,3)$ and perpendicular to the line $5x - 3y - 7 = 0$ in slope-intercept form.

Test – 27

1) A pyramid has a base which is an equilateral triangle. The area of one of its lateral faces is 10cm². Find the total lateral area.

2) The base area of a triangular pyramid is 15 cm², and the total surface area is 60 cm². Find the total lateral area.

3) The length of the bottom edge of a square pyramid is 6 cm. Find the base area of this pyramid?

4) The base area of a square pyramid is 169 cm². Find the length of an edge of the base of the pyramid.

5) What is the base area of a square pyramid with base edge length 2 cm?

6) Find the slant height of a cone with base radius 6 cm and height 8 cm.

7) Find the slant height of a cone with base radius 10 cm and height 20 cm.

8) Find the base circumference of a cone with base radius 3 cm.

9) Find the circumference of the base of a cone with height 8 cm and slant height 10 cm.

10) Find the radius of the base of a cone with height 12 cm and slant height 13 cm.

11) Find the area of the lateral surface of a cone with height 12cm and slant height 13 cm.

12) Find the area of the lateral surface of a cone with height 20cm and slant height 25 cm.

13) What is the surface area of the sphere with radius 7 cm?

14) Find the surface area of a sphere whose largest circle has an area of 16π cm².

15) A circle with radius 10 cm is rotated 360° about its diameter. Find the surface area of the resulting solid.

Test – 28

1) A rectangular prism has sides of length 1 cm and 3 cm, and height 4 cm. Find the surface area of the prism.

2) A rectangular prism has sides of length 2 cm and 3 cm, and height 5 cm. Find volume of the prism.

3) A rectangular prism has a square base with edge length 4 cm and a height of 5 cm. Find the surface area of the prism.

4) A rectangular prism has a square base with edge length 7 cm and a height of 3 cm. Find the surface area of the prism.

5) Find the surface area of a cube with edge length 6 cm.

6) Find the surface area of a cube with edge length 8 cm.

7) Find the volume of a cube with surface area 24 cm².

8) Find the surface area of a cube whose volume is 125 cm³.

9) Find the surface area of a cube whose volume is 27 cm³.

10) Find the surface area of a cube whose volume is 64 cm³.

11) Two cubes have edge lengths that are in the ratio 2:3. Find the ratio of the volumes of the cubes.

12) Two cubes have edge lengths that are in the ratio 7:5. Find the ratio of the surface areas of the cubes.

13) Find the volume of a triangular prism with base area 10cm² and height 6 cm.

14) Find the volume of a triangular prism with base area $4\sqrt{3}$ cm² and height $2\sqrt{3}$ cm.

15) Find the height of a triangular prism whose base area is 12cm² and whose volume is 60 cm³.

Answer Key

Test - 1
1) 104°
2) 160°
3) 15°
4) 105°
5) 150°
6) 290°
7) 20°
8) 165°
9) 80°
10) 81°, 99°
11) 36°, 54°
12) 60°
13) 65°
14) 35°
15) 10°

Test - 2
1) $\sqrt{3}/2$
2) 24
3) 10 cm
4) $5\sqrt{3}$ cm
5) 16 cm
6) $4\sqrt{3}$ cm
7) 25
8) 60 cm
9) $18\sqrt{3}$ cm
10) $\sqrt{76}$ cm
11) 5/3
12) 9/7
13) $6 + 2\sqrt{3}$
14) 40/3
15) 7.5

Test - 3
1) 4/5; 16/25
2) 7.5
3) 80°
4) 24; $16\sqrt{3}$
5) 12 cm
6) 30 cm²
7) 24 cm
8) $8 + 8\sqrt{3}$ cm
9) 48
10) $18\sqrt{3}$; $18+6\sqrt{3}$
11) 112.5; $30+15\sqrt{2}$
12) 120; 60
13) 6 cm
14) $81\sqrt{3}/4$ cm
15) $\sqrt{3}/36$ cm

Test - 4
1) 12
2) 14
3) 20
4) 12
5) 40°
6) 10
7) 150°
8) 18
9) 9
10) 9
11) 144°
12) 12
13) 8
14) 6
15) 24

Test - 5
1) 32 cm
2) 64 cm²
3) $10 + 2\sqrt{21}$
4) $\sqrt{6}$ cm
5) 20 cm
6) 8 cm
7) 75°
8) 24 cm
9) 12 cm²
10) $6\sqrt{3} - 6$
11) 64/3 cm²
12) 64/5
13) $4\sqrt{17}$ cm
14) $12\sqrt{2}$ cm
15) $2\sqrt{3}/3$

Test - 6
1) 375 cm²
2) 96 cm²
3) 72°
4) 40°
5) 52°
6) 108 cm²
7) 68 cm
8) $12\sqrt{3}$ cm
9) 10
10) 960
11) 14 cm
12) 28 cm²
13) 13 cm
14) $10 + 10\sqrt{3}$
15) 192 cm²

Test - 7
1) 48 cm²
2) 36 cm
3) 80 cm²
4) $8\sqrt{13}$ cm
5) 18 cm
6) $2\sqrt{5}$ cm
7) 26 cm
8) 3/7
9) $20\sqrt{3}$
10) $12\sqrt{5}$ cm
11) 384 cm²
12) $4\sqrt{2a+1}$
13) $8\sqrt{19}$
14) 120/13 cm
15) 96 cm²

Test - 8
1) 60 cm²
2) 28 cm
3) 50 cm²
4) 9 cm
5) 76 cm
6) 54 cm²
7) 48 cm
8) 20°
9) 118°
10) 21 cm²
11) 21 cm
12) 40 cm
13) 42 cm
14) 22
15) 44°

Test - 9
1) 14
2) 20π cm
3) 3/4
4) $2\sqrt{2}\pi$ cm
5) $5\pi/2$ cm
6) $10\pi/3$ cm
7) 8π cm
8) $5\sqrt{2}$ cm
9) 80°
10) 40°
11) 40°
12) 180
13) 150°
14) 80°
15) 6

Test - 10
1) 14 cm
2) 1 cm
3) 9 cm
4) 9/25
5) 18 cm
6) $200\pi/9$ cm²
7) 12π cm²
8) 30°
9) $32\pi/3$ cm²
10) $8\pi - 16$ cm²
11) 14 cm
12) 4
13) 3π cm²
14) 36%
15) 96%

Test - 11
1) 40°
2) 150°
3) 20°
4) 130°
5) 136°
6) 12
7) 3
8) 4
9) $2\sqrt{31}$
10) $18 + 6\sqrt{3}$
11) $15+5\sqrt{3}$ cm
12) $4\sqrt{2}$
13) $\sqrt{21}$
14) 12
15) 3

Test - 12
1) $14+4\sqrt{7}+2\sqrt{21}$
2) $16\sqrt{3}$
3) 25/4
4) 36/7
5) 5
6) 16 cm²
7) 18°
8) 90°
9) 15°
10) 44°
11) 30°
12) 140°
13) 80°
14) 12°
15) 40°

Test - 13
1) $125°$
2) 12
3) $4\sqrt{2}$ cm
4) $5\sqrt{2}$ cm
5) $25/2$ cm²
6) 8 cm
7) $\sqrt{31}$
8) $\sqrt{15}$
9) 5 cm
10) $10\sqrt{5}$
11) $6\sqrt{3}$ cm
12) $65°$
13) $5\sqrt{3}$ cm
14) 12
15) $8\sqrt{3}$ cm

Test - 14
1) $\sqrt{3}$ cm
2) $9+3\sqrt{3}+3\sqrt{6}$
3) $2\sqrt{5}$ cm
4) $20°$
5) $105°$
6) $30°$
7) $22.5°$
8) $80°$
9) $168°$
10) $24°$
11) $100°$
12) $50°$
13) $45°$
14) $15°$
15) $80°$

Test - 15
1) $25\sqrt{3}/4$
2) $36\sqrt{3}$
3) 48
4) 84 cm²
5) 15 cm²
6) 20 cm²
7) 18 cm²
8) 1 cm²
9) $9/5$
10) $81\sqrt{3}/4$
11) 18 cm²
12) 32 cm²
13) $\sqrt{3}/8$ cm²
14) $6\sqrt{3}$ cm²
15) 60 cm²

Test - 16
1) $720°$
2) $1440°$
3) 20
4) 35
5) 12
6) 14
7) 40
8) $45°$
9) 50 cm
10) 18
11) 36
12) $2\sqrt{14}$
13) $49/16$ cm²
14) $4\sqrt{3}/3$
15) $4\sqrt{6}$ cm

Test - 17
1) 72 cm²
2) 25 cm²
3) $\sqrt{146}$ cm
4) 8 cm
5) $60°$
6) $150°$
7) 24 cm²
8) $4\sqrt{3}-4$
9) 54 cm²
10) $2\sqrt{26}$ cm
11) 12
12) $\sqrt{2}/4$ cm
13) $4\sqrt{2}$ cm
14) $2/5$
15) $\sqrt{6}$ cm

Test - 18
1) $2\sqrt{13}$ cm
2) 150 cm²
3) $10\sqrt{3}$ cm
4) $6\sqrt{10}$ cm²
5) 144 cm²
6) $27\sqrt{2}$ cm²
7) $72°$
8) $9\sqrt{3}$ cm
9) 48 cm²
10) $4+4\sqrt{3}$ cm
11) 160 cm²
12) $40°$
13) $32°$
14) 10 cm
15) 20 cm

Test - 19

1) 12

2) $2\sqrt{6}$ cm

3) 169 cm²

4) $27\sqrt{21}$ cm²

5) 34

6) $48\sqrt{3}$ cm²

7) 9 cm

8) $7 + 3\sqrt{3}$ cm

9) 126 cm²

10) $125\sqrt{3}/2$ cm²

11) 100°

12) 4

13) $22\sqrt{3}$ cm

14) 224 cm²

15) $64\sqrt{3}$ cm²

Test - 20

1) 48

2) 20 cm

3) $10\sqrt{2}$ cm

4) $8\sqrt{34}$

5) 52 cm

6) 18 cm

7) $2\sqrt{39}$ cm

8) 600 cm²

9) 25 cm²

10) $4\sqrt{41}$ cm

11) $\sqrt{13}$ cm

12) 17 cm

13) $4\sqrt{3}$

14) 60 cm

15) 120/13 cm

Test - 21

1) 24 cm²

2) 55°

3) $16\sqrt{2}$ cm²

4) $3\sqrt{39}$

5) 32°

6) 37°

7) 30°

8) 80°

9) 25°

10) 60°

11) 88°

12) 42°

13) 48°

14) 22°

15) 30°

Test - 22

1) 50°

2) 96°

3) 38°

4) $4\pi/3$

5) $16\pi/3$

6) 165°

7) 34°

8) 70°

9) 68°

10) 38°

11) 32°

12) 66°

13) 128°

14) 26°

15) 111°

Test - 23

1) 15 cm

2) 26 cm

3) $25\pi/4$ cm²

4) $25\pi - 50$ cm

5) $16\pi - 32$ cm

6) 7.2π cm

7) $2\sqrt{2}$ cm

8) 4π cm

9) $9\pi/2 - 9$ cm

10) 196π cm²

11) 2 cm

12) 40 cm

13) $400\sqrt{3}/3 - 400\pi/9$

14) $16\sqrt{3}$ cm²

15) 16/49

Test - 24

1) 53°

2) 136°

3) 140°

4) 154°

5) 18

6) 3

7) 3

8) $5\sqrt{10}$

9) $22 + 11\sqrt{2}$

10) $18 + 6\sqrt{3}$

11) $2\sqrt{15}$ cm

12) 20 cm

13) $4\sqrt{7} + 2\sqrt{21} + 14$

14) 7 cm

15) 56/9

Test - 25
1) 90/7
2) 120°
3) 12°
4) 50°
5) 36°
6) 139°
7) 92°
8) 60/7
9) 42°
10) 122°
11) 12 cm
12) $2\sqrt{2}$ cm
13) $6\sqrt{2}$ cm
14) $16 + 8\sqrt{2}$ cm
15) $\sqrt{95}$ cm

Test - 26
1) $\sqrt{42}$
2) $\sqrt{10}$
3) $6\sqrt{2}$ cm
4) $2\sqrt{11}$
5) $8+4\sqrt{3}$ cm
6) 66°
7) $12 + 12\sqrt{3}$
8) $16 + 8\sqrt{3}$
9) $11\sqrt{7}$ cm
10) $7 + \sqrt{7}$ cm
11) $y-6 = 4(x-3)$
12) $y = x-5$
13) $y = -\frac{3}{2}x+9$
14) $y = -\frac{6}{5}x$
15) $y = -\frac{3}{5}x$

Test - 27
1) 30 cm²
2) 45 cm²
3) 36 cm²
4) 13 cm
5) 4 cm²
6) 10 cm
7) $10\sqrt{5}$ cm
8) 6π cm
9) 12π cm
10) 5 cm
11) 65π cm²
12) 375π cm²
13) 196π cm²
14) 64π cm²
15) 400π cm²

Test - 28
1) 38 cm²
2) 30 cm³
3) 112 cm²
4) 182 cm²
5) 216 cm²
6) 384 cm²
7) 8 cm³
8) 150 cm²
9) 54 cm²
10) 96 cm²
11) 8/27
12) 49/25
13) 60 cm³
14) 24 cm³
15) 5 cm

About the Authors

Tayyip Oral graduated from Qafqaz University in Azerbaijan in 1998 with a Bachelor's degree in Engineering, and he received an MBA from the same university in 2010. Tayyip is an educator who has written several books related to math and intelligence questions. He has previously taught math and IQ classes at Baku Araz preparatory school for 13 years.

Serife Turan graduated from Mehmet Akif Ersoy University in 2008 with a B.A. in education , and she worked for three years as a math teacher. She is currently pursuing her Master's Degree in the School of Education and Human Development at the University of Houston Victory.

Dr. Steve Warner earned his Ph.D. at Rutgers University in Mathematics, and he currently works as an Associate Professor at Hofstra University. Dr. Warner has over 15 years of experience in general math tutoring and over 10 years of experience in SAT math tutoring. He has tutored students both individually and in group settings and has published several math prep books for the SAT, ACT and AP Calculus exams.

BOOKS BY TAYYIP ORAL

1. T. Oral and Dr. S. Warner, 555 Math IQ Questions for Middle School Students, USA, 2015

2. T. Oral and Sevket Oral, 555 Math IQ Questions for Elementary School Students, USA, 2015

3. T. Oral, IQ Intelligence Questions for Middle and High School Students, USA, 2014

4. T. Oral, E. Seyidzade, Araz publishing, Master's Degree Program Preparation (IQ), Cag Ogretim, Araz Courses, Baku, Azerbaijan, 2010, Azerbaijan.

5. T. oral,M. Aranli, F. Sadigov, and N. Resullu, A Text Book for Job Placement Exam in Azerbaijan for Undergraduate and Post Undergraduate Students in Azerbaijan, Resullu publishing, Baku, Azerbaijan - 2012 (3.edition)

6. T. Oral and I. Hesenov, Algebra (Text Book), Nurlar Printing and Publishing, Baku, Azerbaijan, 2001.

7. T. Oral, I. Hesenov, S. Maharramov, and J. Mikaylov, Geometry (Text Book), Nurlar Printing and Publishing, Baku, Azerbaijan, 2002.

8. T. Oral, I. Hesenov, and S. Maharramov, Geometry Formulas (Text Book), Araz courses, Baku, Azerbaijan, 2003.

9. T. Oral, I. Hesenov, and S. Maharramov, Algebra Formulas (Text Book), Araz courses, Baku, Azerbaijan, 2000

BOOKS BY DR. STEVE WARNER

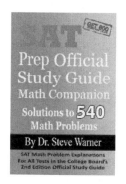
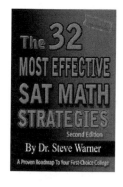

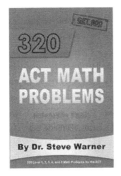

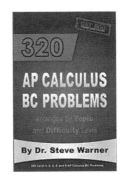

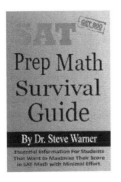

24655160R00107

Made in the USA
Middletown, DE
01 October 2015